The Era of Internet of Things

Khaled Salah Mohamed

The Era of Internet of Things

Towards a Smart World

 Springer

Khaled Salah Mohamed
Mentor, A Siemens Business
Cairo, Egypt

ISBN 978-3-030-18135-2   ISBN 978-3-030-18133-8   (eBook)
https://doi.org/10.1007/978-3-030-18133-8

This Springer imprint is published by the registered company Springer Nature Switzerland AG
The registered company address is: Gewerbestrasse 11, 6330 Cham, Switzerland

To my beloved daughters

Preface

Internet of things (IoT) has entered its golden age. This book presents a smart introduction and guide to the IoT. It discusses all the necessary components and knowledge to start being a vital part of the IoT revolution. IoT is all about intelligence, not just control. Now, IoT is a fast-changing set of technologies and architectures. In this book, we learn how to create smart-IoT solutions to help solve different problems. We present the most important aspects of IoT, the various applications of IoT, and the enabling technologies for IoT. This book presents main IoT concepts and abstractions explaining many case studies such as smart homes, smart agriculture, and smart automotive. Analysis of IoT strength, weakness, opportunities, and threats of IoT is also presented.

Cairo, Egypt   Khaled Salah Mohamed

Contents

    Communication and Networking 52
    4.1   Personal Area Network (PAN). 53
    4.2   Local Area Network (LAN). 57
    4.3   Wide Area Network (WAN) 58
    4.4   Broadcast Network (BN). 61
    4.5   Global Network (GN) 62
5   IoT Internet Layer. 62
    5.1   IPV6/6LowWPAN. 62
6   IoT Application Layer 63
    6.1   CoAP. 63
    6.2   MQTT 63
7   Comparison between Different IoT Protocols 64
8   The Future of Wireless Technology 64
9   Conclusions 66
References. 66

IoT Cloud Computing, Storage, and Data Analytics 71
1   Introduction 71
2   Cloud Computing 72
    2.1   Cloud Computing: What? 72
    2.2   Cloud Computing: Why?. 72
    2.3   Cloud Computing: How?. 73
3   Edge/Fog Computing 81
4   Data Analytics for Big Data: Machine Learning 83
    4.1   IoT Analytics: Why? 85
    4.2   The IoT Edge Data Analytics: Real Cases 86
    4.3   IoT Data Analytics: Types. 89
5   Conclusions 89
References. 89

IoT Application Layer: Case Studies and Real Applications 93
1   Introduction 93
2   IoT Case Studies. 94
    2.1   Hospital Model/e-Health. 95
    2.2   Museum Model 96
    2.3   Inventory Model 96
    2.4   Advertising Model. 96
    2.5   Food Tracing Model 96
    2.6   Residence Model. 97
    2.7   Maintenance Model. 97
    2.8   Fire Alarm Model 97
    2.9   Attendance Model. 97
    2.10 Access Control Model. 97
    2.11 Library Model 98
    2.12 Cashless Payment Model. 98

About the Author

Khaled Salah Mohamed received his B.Sc. degree in Electronics and Communications Engineering with distinction and honors degree in 2003 from Ain Shams University, Cairo, Egypt. He received his M.Sc. and his Ph.D. degrees in Electronics and Communications in 2008 and 2012, respectively. He received his M.B.A. degree in 2016. He joined Mentor Graphics Corporation, where he designed many SoC IPs such as AHB, HDMI, HDCP, eMMC, SDcard, HMC, and LPDDR5. Currently, Dr. Khaled Salah is an Engineering Lead at the Emulation Division at Mentor Graphics, Egypt. Dr. Khaled Salah has published three books and more than 93 research papers in the top refereed journals and conferences. His research interests are in 3D integration, IP Modeling, Internet of Things, artificial intelligence, and SoC design. He is a senior IEEE member. Dr. Khaled served as a reviewer for several conferences and journals, including IEEE Transactions on VLSI, IEEE Transactions on Circuits and Systems II, IEEE Transactions on Semiconductor Manufacturing, IEEE Microwave and Wireless Components Letters, IEEE Transactions on Microwave Theory and Techniques, and ELSEVIER Microelectronics Journal.

The Era of Internet of Things: Towards a Smart World

1 Introduction

Internet of Things (IoT) is expected to revolutionize our lives. IoT is now a growing industry. Analysts predict that IoT products and services will grow exponentially in next years. By 2020, it is expected that more than 50 billion IoT devices will be connected as depicted in Fig. 1 [1]. It is a confluence of different sectors: embedded systems, communication systems, sensors/actuators, WWW, and mobile applications. However, IoT is still having many challenges and limitations due to a number of factors, which limit the full exploitation of the IoT. The concept of IoT was born in 1999 by **Kevin Ashton** in the United States. Global Internet of Things market set to reach $318bn by 2023 [2].

In this chapter, we present a comprehensive survey of IoT and a strength-weakness-opportunities-threats (SWOT) analysis for it. Moreover, this chapter provides a historical perspective with special emphasis on recent works and future perspective.

1.1 IoT: What?

There is no single definition for Internet of Things (IoT). IoT is a new dimension of the internet and a new generation of services. IoT means anything can communicate with anything in any place at any time using any protocols as depicted in Fig. 2. Because information is sent to the internet from any place, so you can access them from any place [4]. IoT is like a human society, with minimum human intervention as things have virtual identities to be known. IoT will enable "smart X," where "X" can be anything such as TV, watch, glass, clock, coffee machine, and car. History of internet of things "IoT" back to 1997, but the first conference was launched on 2008. IoT AKA machine to machine "M2M," device to device "D2D," and "Ubiquitous."

© Springer Nature Switzerland AG 2019
K. S. Mohamed, *The Era of Internet of Things*,
https://doi.org/10.1007/978-3-030-18133-8_1

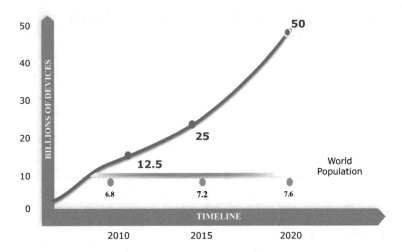

Fig. 1 IoT timeplan: Growth of things connected to the Internet

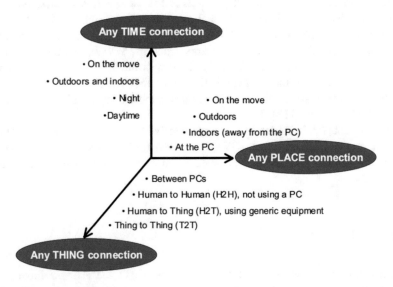

Fig. 2 IoT smartness [3]

In IoT, we start with a "Thing" and add computational intelligence to improve its function, then add a network connection to further enhance its function as depicted in Fig. 3. Figure 4 shows cloud computing as an enabling technology for IoT [5]. Microsoft Azure is an example for a cloud computing platform. Cloud computing means that devices exchange information through a cloud infrastructure. Evolution of IoT is shown in Fig. 5. Things in IoT can be physical or virtual as depicted in Table 1.

1.2 IoT: Why?

By enabling IoT, control of daily life in an intelligent and easier way is feasible. There are many reasons that make IoT feasible such as Internet infrastructure already exists and internet available almost everywhere in the developed world. Moreover, hardware size allows incorporation into a device. Besides, cost of hardware has decreased. IPv6 protocol has large address space, so we can assign an IP for each thing on the earth. There is infinite number of applications for IoT such as traceability, smart home, smart office, smart campus, smart shopping, and smart

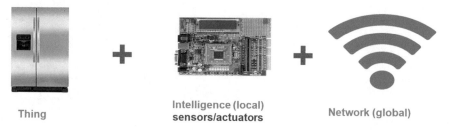

Fig. 3 IoT: "Thing" and add computational intelligence to improve its function then add a network connection to further enhance its function

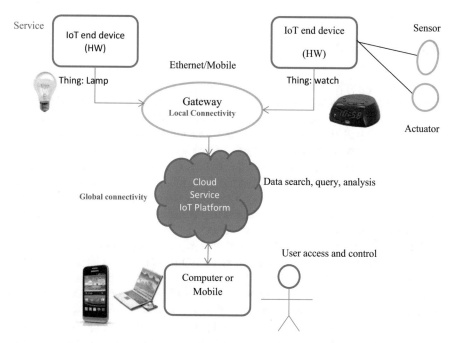

Fig. 4 The model explains how such IoT architecture works: A device with sensors and actuators allows a direct, physical interaction. Since it is also connected to a server via the Internet, it can interact with it virtually, using a browser or app—locally, or remotely

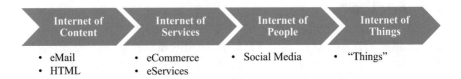

Fig. 5 Evolution of IoT [6]. The Internet services have evolved from conventional point-to-point data exchange, world wide web (WWW), mobile and social applications, to the recent IoT services

Table 1 Physical and virtual things

Physical things	Virtual things
Car	Email
Energy	Twitter
People	Database storage
Pets	Stocks
Temperature	Weather forecasting
Weight	Facebook

clothes. Table 2 and Fig. 6 show different IoT usage and applications. IoT will not only control our homes, but also our business, society, cities, and our lives [7].

1.3 IoT: How?

IoT is not the result of a single novel technology; instead, several complementary technical developments provide capabilities that are taken together to help bridge the gap between the virtual and physical world. In IoT concept, everything that can be automated will be automated. IoT is about intelligence not just control. IoT enabling technologies basically consist of four main functions: sensing, communication, control, and actuators which have a great analogy with human body as depicted in Fig. 7. A large number of industrial data, usually referred to big data, are collected from IoT. The IoT concept has been designed to perform several main distinctive actions: collect data, transfer/transmit/exchange data, change/process data, store data, and personalize/execute data.

Sensors can be real sensors or virtual sensors to collect data from the internet. Communication can be done using many types of protocols such as RFID, AD-HOC, Ethernet, Wi-Fi, 3G, 4G, Bluetooth, ZigBee, USB, WSN, and IPv6, which are ranging from short-range to long-range communications (Table 3). For example, Bluetooth for short-range connectivity; Wi-Fi for medium scale connectivity; cellular technologies for large scale connectivity. Control is done using FPGA, ASIC, or processors. Actuators examples are motor, alarm, and oven. IoT architecture consists of three layers: physical layer, communication layer, and application layer as depicted in Fig. 8.

Table 2 Domain of IOT usage: applications

Smart Homes [8]	– Control and home security [9–13] – Intelligent systems maintenance – Intelligent heating and cooling systems – Control and monitoring of energy consumption (water, electricity, gas) – Facial and biomedical recognition
Smart Cities	– Intelligent monitoring – Automatic transport – The exact energy management systems – Environmental monitoring
Smart Transportation/ Automotive	– Intelligent traffic control systems – Intelligent systems for maintenance of roads (land, air and sea) – Intelligent Systems Parking – RFID tags communication.
Smart Retail and logistics	– Supply Chain Control – Intelligent Shopping Applications – Smart Product Management – Inventory tracking – Point-of-sale terminals – Vending machines
Smart Agriculture	– Sensors check the soil moisture and temperature: Soil Moisture Management – Smart Irrigation – Smart dust.
Smart Factories and Industries/ Business	– Indoor Air Quality – Temperature Monitoring – Ozone Presence – Indoor Location – Vehicle Auto-diagnosis – Sensors check the soil moisture and temperature.
Smart Health Care	– Patients Surveillance – Sportsmen Care – Ultraviolet Radiation – Smart hospitals.
Smart Wearable	– Smart Glasses – Smart clothes – Sleep Sensor – Smart watch.
Others	– Smart museums – Smart schools – ATMs.

IoT connectivity layers are shown in Fig. 9. Nodes connected to each other using LANs which may or may not be connected to the internet (WAN) through gateways (using proxy to connect to the internet or without connectivity to provide intra-connectivity between different LANs). Devices/nodes are often connected to a gateway in cases when the device is not capable of directly connecting to further systems, e.g., if the device cannot communicate via a particular protocol or because

Fig. 6 Smartness domains

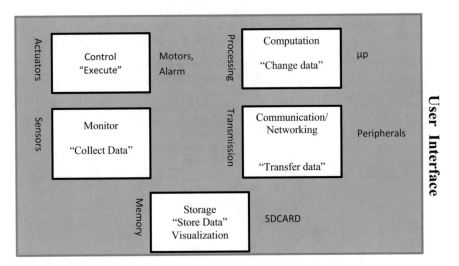

Fig. 7 General architecture for IoT platform. IoT describes the connection of devices with embedded sensors, actuators, and software by networking technologies

of other technical limitations. To solve these problems, a gateway is used to compensate such limitations by providing required technologies and functionalities to translate between different protocols and by forwarding communication between devices and other systems. A gateway is, therefore, responsible for supporting the required communication technologies and protocols in both directions and for translating data if necessary. For instance, a device communicates with a gateway via an IoT protocol, such as ZigBee or MQTT. When the gateway receives a message in a proprietary binary format from the device, the gateway translates the information into **JSON** or **XML** and forwards the data to a system in the world wide web. Likewise, the gateway may translate commands into communication technologies, protocols, and formats supported by the respective device. The gateway may already execute some data processing functions, such as data aggregation, depending on its processing capabilities.

To implement an IoT algorithm, you have many options from software and hardware point of view. Hardware options for IoT are shown in Table 4. Software options can be: Python, embedded C, Java, Javascript. There is no single architecture for IoT.

Table 3 Different communication technologies for IOT

Technology	Frequency	Range	Data rate
Bluetooth	2.4GHz	50–150 m	1Mbps
ZigBee	2.4GHz	10–100 m	250 kbps
Wi-Fi	2.4GHz	~50 m	600 mbps
NFC	13.56 MHz	10 cm	100–420 kbps

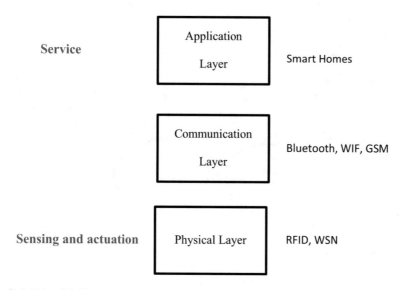

Fig. 8 IoT simplified layers. IoT architecture consists of three layers: physical layer, communication layer, and application layer, containing a set of IoT protocols

1.4   IoT: When?

We can make full use of IoT technology when we overcome all its challenges and limitations. Any IoT system should satisfy 4s's rule: simple, secure, smart, and scalable. Security and privacy are a challenge in IoT. Large number of IoT devices means increased threats, so a new security level is needed. We need to protect the cloud, the communication, and ensure privacy and integrity. In some applications, we need real-time processing and maybe novel simulation techniques. Sensor reliability is an important limitation. There is no unified protocols and standardization for IoT. We need regulations to avoid multiple identities. Scalability is a challenge in IoT when we have massive number of devices. Low power and power harvesting is very important in IoT as most devices are battery-based devices [15].

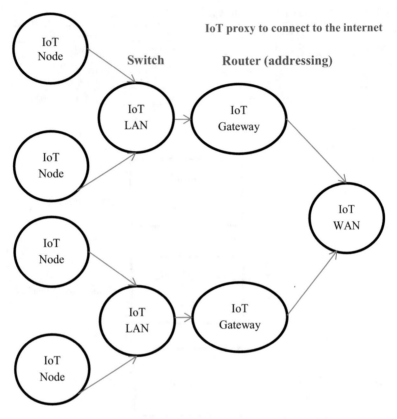

Fig. 9 Connectivity layers. Nodes connected to each other using LANs which may or may not be connected to the internet (WAN) through gateways (using proxy to connect to the internet or without connectivity to provide intra-connectivity between different LANs)

Table 4 Hardware options for IoT

Technology	Advantages	Disadvantages
A general-purpose microprocessor	• A short development cycle • Support for a variety of high-level languages such as Java and C++	• Serial execution • Static hardware configuration • To increase performance, you need to increase clock speeds which increase energy consumption
Specialized processors such as DSPs and GPUs	• Very efficient at implementing specific tasks such as multiply/accumulate cycles	• Not efficient for all applications [14]
ASIC	• Superior performance, area, and power efficiency	• Long development cycles • Zero flexibility once fabricated
FPGA	• Flexibility	• Cost

1.5   IoT: Requirements/Characteristics

- Minimal human intervention during operation or configuration.
- Long battery life time as most of IoT devices are battery-operated devices.
- Security and privacy as IoT devices cannot afford resource-demanding encryption protocols [16, 17].
- Ensuring quality of service (QoS) and efficient communication.
- Heterogeneity: The devices in the IoT framework are heterogeneous as they are based on different platforms and networks. They can interact with other devices or service platforms through different networks.

1.6   IoT: Challenges

- Integration of hardware and software from several vendors.
- Interaction with devices using multiple wireless protocols.
- Real-time data collection and analytics.
- Seamless and secure connection to cloud.
- Cost of design and deployment.
- Remote device management and diagnosis.
- IoT CAD development tools.
- Sensor reliability: finite life time.
- Unified standards.
- Large number of connection of nodes.
- Powering billions of connected devices.

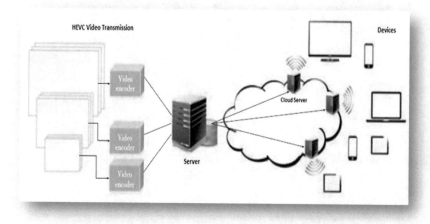

Fig. 10   HEVC encoding

- Wireless communication with less power: Video encoding such as HEVC is an example of compression for low-power transmission as shown in Fig. 10.
- Bandwidth management.
- Scalability: up-scaling and down-scaling.
- Security and privacy: Security goals of IoT protocols are summarized in Fig. 11. In the past, IT security has been based on establishing secure boundaries and firewalls around internal IT systems. The IoT model is defined by extreme access to many different devices that collect and leverage vast amounts of data. The concept of controlled access is changing with the IoT model to one of controlled trust to enable the wide range of possible solutions. IoT implementations must effectively deal with authorization, authentication, access control, privacy, and trust requirements while not negatively impacting usability objectives [18].
- Quality of service (QoS) [19].
- Building a general framework of IoT is very complex task because of heterogeneity in devices, technologies, platforms, and services, operating in the same system.

1.7   Enabling/Key Technologies for IoT

- **Internet infrastructure** already exists and number of internet users is huge.
- **Cloud computing**: With millions of devices expected to come by 2020, the cloud seems to be the only technology that can analyze and store all the data effectively. It analyzes the useful information obtained from the sensors and even provides good storage capacity.

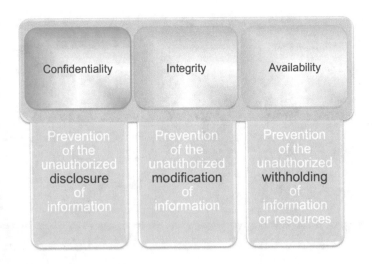

Fig. 11 Security goals of IoT protocols

- **IPv6:** Address space can assign an IP for each thing on the earth. It supports addresses up to 2^{128}.
- **Cost** of hardware has decreased (Fig. 12).
- **Hardware size** allows incorporation into a device (Fig. 13).
- **Computational efforts** increased, this was impossible with old machines.
- **RFID**: RFID is the key technology for making the objects uniquely identifiable. RFID system is composed of readers and associated RFID tags which emit the identification, location, or any other specifics about the object, on getting triggered by the generation of any appropriate signal.
- **5G:** The 5G networks are expected to massively expand today's IoT that can boost cellular operations, IoT security, and network challenges and driving IoT future to the edge.
- **WSN:** Wireless Sensor Network (WSN) is low-cost, low-power miniature devices for use in remote sensing applications.

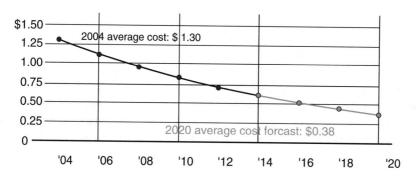

Fig. 12 The average cost of IoT sensors is falling

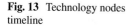

Fig. 13 Technology nodes timeline

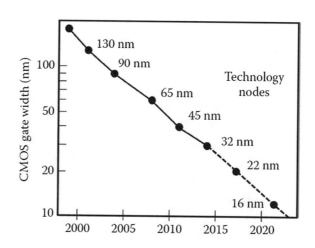

- **Machine learning.**
- **Big data**.
- Micro-Electro-Mechanical Systems (**MEMS**): The application of MEMS sensors to the IoT-enabled markets will require sensors to shrink further and to work even more power-efficient as in smartphones.
- **Low-power embedded systems.**
- **Smart networks.**
- **Nanotechnology.**
- **Charging Technologies:** To support the power requirements of IoT devices and components, wireless charging technologies are becoming increasingly important. Inductive charging uses electromagnetic fields to transfer power between devices such as a smartphone and a charging mat or pad that are in direct contact with each other, while resonance charging uses magnetic fields to transfer power between devices. In addition to eliminating the need for power cords or cables, both charging technologies allow IoT devices to be constructed without openings or sockets for power cords, making them less susceptible to damage from exposure to water and other liquids.

1.8   IoT: An Example

An IoT example is shown in Fig. 14. Sensors can monitor and track any changes. Arduino as a controller can analyze and take decisions which can be sent using mobile 4G communications from place1 to place2, where another Arduino controllers can analyze the commands and send actions to the actuators [20, 21].

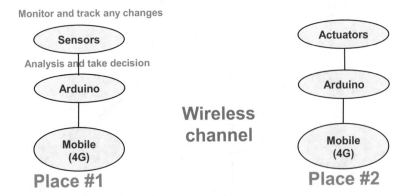

Fig. 14 An IoT example

2   SWOT Analysis of IoT

In this section, we are providing a SWOT (strength, weakness, opportunities, threats) analysis for IoT [22–36]:

2.1   Strength of IoT

- Portability.
- Scalability: real-time connectivity of billions of devices.
- Unlimited functionality.

2.2   Weakness of IoT

- Dependency on Internet (network outage).
- Dependency on electricity (electricity outage).
- Security.

2.3   Opportunities of IoT

- A new field for a startup company.
- Business consulting.

2.4   Threats of IoT

- Social isolation.
- Dependency on machines may reduce human abilities.
- Regulations.

3   IoT Testing

The following types of testing need to be performed within an IoT ecosystem [37–40]:

- **Functional testing**: which validates the correct functionality of the IoT application?

- **Connectivity testing**: it is responsible for testing the wireless signal in order to determine what happens in case of weak connection, or when there are many devices trying to communicate.
- **Performance testing**: which validates the communication and computation capabilities? Stress testing can be used in order to find how many simultaneous connections can be supported by a specific device.
- **Security testing**: focus in privacy, authorization and authentication features.
- **Compatibility testing**: verifies the correct functionality under different protocols and configurations.
- **Exploratory testing**: also called user experience tests.

4 New Trends in IoT

4.1 Sensors as a Service

Sell sensor data [41]. Sensing-as-a-Service (SEaaS) enables the exchange of non-proprietary data to truly connect the Internet of Things (IoT). The delivery of sensor data either on demand or via a data stream. At the end of the day, the consumers of data from IoT sensors are interested in the data points, not necessarily the ins and outs of where they came from. SEaaS is a vision and a business model that promotes data exchange between data owners and data consumers [42]. The sensing-as-a-service model consists of four conceptual layers: (1) sensors and sensor owners, (2) sensor publishers, (3) extended service providers, and (4) sensor data consumers as depicted in Fig. 15.

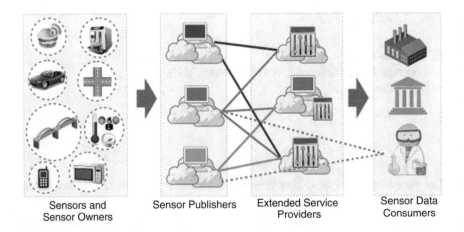

| Sensors and Sensor Owners | Sensor Publishers | Extended Service Providers | Sensor Data Consumers |

Fig. 15 The sensing-as-a-service model [43]

4.2   Digital Twins

Each physical system has a digital simulation twin that can simulate real-time sensor data that enters to the physical system and generates recommendations to improve the performance at real time, i.e., it is the ability to make a virtual representation of the physical elements [44]. Figure 16 shows the evolution of IoT till the forthcoming tactile Internet which will facilitate the integration between digital sphere and our physical environments, covering advanced use cases of machine-to-machine (M2M) communication. Digital twins are built on simulation modeling—combining a simulation model with data from its real-world counterpart. Digital twins can be used for diagnostics, forecasting, and visualization.

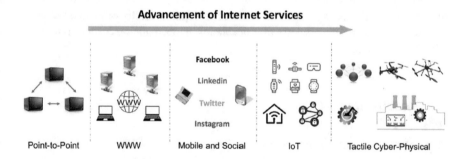

Fig. 16   The evolution of IoT [45]

Fig. 17   How blockchain works

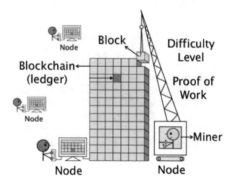

4.3 Managing IoT Devices Using Blockchain Platform

Since the start of **Bitcoin** in 2008, blockchain technology emerged as the next revolutionary technology. Though blockchain started off as a core technology of Bitcoin, its use cases are expanding to many other areas including finances, Internet of Things (IoT), and security [46, 47]. A blockchain is a continuously growing list of records, called blocks, which are linked and secured using cryptography as shown in Fig. 17 [48–50].

A blockchain is a basic data structure first proposed by Satoshi Nakamoto in 2008 for the peer-to-peer currency known as Bitcoin. A blockchain is composed of many blocks, which can contain any type of data, though they are most often used to keep a record of various transactions between peers. These blocks are linked together backwards, and each block verifies the integrity of its previous block through its hash. Tampering with a previous block will invalidate its hash, making it easily noticeable. Calculating a new hash also known as mining is a very demanding process, and the modification of one block has an effect on every younger block linked to it. While mining is very difficult, verifying the validity of a mined block is very easy for peers. This property of blockchains deters malicious users from modifying block data [51].

5   Conclusions

This chapter presents a comprehensive survey of IoT and a SWOT analysis for it. Moreover, it presents the different fundamental and basic concepts for IoT and its overall operation. This chapter presents the different components needed to build an IoT system and explains its different layers. Moreover, pros and cons of IoT are analyzed. Moreover, we introduce some key industrial and consumer applications of IoT.

References

1. Restuccia, F., D'Oro, S., & Melodia, T. (2018). Securing the internet of things: New perspectives and research challenges. *IEEE Internet of Things Journal, 1*(1).
2. Retrieved from https://www.irishtimes.com/business/technology/global-internet-of-things-market-set-to-reach-318bn-by-2023-1.3705819.
3. Recommendation ITU-T Y. *2060: Overview of the Internet of things.* Retrieved from http://www.itu.int/rec/T-REC-Y.2060-201206-I.
4. Yaqoob, I., Ahmed, E., Hashem, I. A. T., Ahmed, A. I. A., Gani, A., Imran, M., & Guizani, M. (2017). Internet of things architecture: Recent advances, taxonomy, requirements, and open challenges. *IEEE Wireless Communications, 24*(3), 10–16.

5. Tayyaba, S. K., Shah, M. A., Khan, O. A., Ahmed, A. W. (2017). Software defined network (SDN) based internet of things (IoT): A road ahead. *Proceedings of the International Conference on Future Networks and Distributed Systems*. ACM, pp. 1–8.
6. Jadoul, M. (2015). *The IoT: The next step in internet evolution*. Retrieved March 11, 2015, from http://www2.alcatel-lucent.com/techzine/iot-internet-of-things-next-step-evolution/.
7. Sethi, P., & Sarangi, S. R. (2017). Internet of things: Architecture, protocols, and applications. *Journal of Electrical and Computer Engineering, 2017*(9324035), 25.
8. Dorri, A., Kanhere, S. S., Jurdak, R., Gauravaram, P. Blockchain for IoT security and privacy: The case study of a smart home. *Proceedings of the IEEE International Conference on Pervasive Computing and Communications (PerCom) Workshops*. IEEE, March 2017, pp. 618–623.
9. Retrieved from http://www.computerweekly.com/feature/The-internet-of-things-is-set-to-change-security-priorities
10. Retrieved from http://www.computerweekly.com/news/2240209213/APIs-key-to-security-of-internet-of-things-says-Axway
11. Retrieved from http://www.infineon.com/cms/en/applications/chip-card-security/internet-of-things-security/
12. Retrieved from http://internetofthingsagenda.techtarget.com/definition/IoT-security-Internet-of-Things-security
13. Retrieved from http://www.securerf.com/solutions/
14. Salah, K., AbdelSalam, M. (2017). A comparative analysis between FPGA and GPU for solving large numbers of linear equations. *Microelectronics (ICM), 2017 29th international conference on IEEE*.
15. Shafique, K., Khwaja, B. A., Khurram, M. D., Sibtain, S. M., Siddiqui, Y., Mustaqim, M., Chattha, H. T., & Yang, X. (2018). Energy harvesting using a low-cost Rectenna for internet of things (IoT) applications. *IEEE Access, 6*, 30932–30941.
16. Burg, A., Chattopadhyay, A., & Lam, K. Y. (2018). Wireless communication and security issues for cyber-physical systems and the internet-of-things. *Proceedings of the IEEE, 106*(1), 38–60.
17. Ammar, M., Russello, G., & Crispo, B. (2018). Internet of things: A survey on the security of IoT frameworks. *Journal of Information Security and Applications, 38*, 8–27.
18. Riahi Sfar, A., Natalizio, E., Challal, Y., & Chtourou, Z. (2018). A roadmap for security challenges in the internet of things. *Digit. Commun. Netw., 4*(2), 118–137. https://doi.org/10.1016/j.dcan.2017.04.003.
19. Ferrari, P., Flammini, A., Rinaldi, S., Sisinni, E., Maffei, D., & Malara, M. (2018). Impact of quality of service on cloud based industrial IoT applications with OPC UA. *Electronics, 7*, 109.
20. Hanes, D., Salgueiro, G. (2017). *IoT fundamentals: Networking technologies, protocols, and use cases for the internet of things*.
21. Kranz, M. (2016). *Building the internet of things: Implement new business models, disrupt competitors, transform your industry*.
22. Dhanalaxmi, B., Apparao Naidu, G. (2017). A survey on design and analysis of robust IoT architecture. *ICIMIA*.
23. Milan Zdravkovi_c, MiroslavTrajanovi_c, Jo~aoSarraipa, Ricardo Jardim-Gon_calves and MarioLezoche. Survey of internet-of-things platforms. *6th International Conference on Information Society and Technology, ICIST 2016, Kopaonik, Serbia*, pp. 216–220, February 2016.
24. Abed, A. A. Internet of things (IoT): Architecture and design. *Al-Sadeq International Conference on Multidisciplinary in IT and Communication Science and Applications*, IRAQ, May 2016.

25. Al-Qaseemi, S. A., Almulhim, H. A., Almulhim, M. F., Chaudhry, S. R. *IoT architecture challenges and issues: Lack of standardization*. Future Technologies Conference, San Francisco, United States, December, 2016.
26. Dell Enters Embedded PC Market with New Embedded Box PCs, Helping Smart Systems Connect to the Internet of Things. [Online].
27. Retrieved   September   10,   2017,   from   http://www.businesswire.com/news/home/20160222006280/en/Dell-Enters-Embedded-PC-Market-New-EmbeddedDellIoTArchitecture
28. Alvarez, A. (2017). Microsoft announces IoT central: SaaS platform to simplify the internet-of-things. [Online]. Retrieved September 10, 2017, from https://www.starwindsoftware.com/blog/microsoft-announces-iot-central-saas-platform-to-simplify-the-internet-of-things.
29. Architecture: Real Time Stream Processing - Internet of Things. [Online]. Retrieved September 10, 2017, from http://ec2-54-66-129-240.ap-southeast2.compute.amazonaws.com/httrack/docs/cloud.google.com/solutions/architecture/streamprocessing.html.
30. Tyagi, N. (2016). A Reference Architecture for IoT. *International Journal of Computer Engineering and Applications, X*(I).
31. Jose, A., Prasanthkumar, P. V., & Mithun, T. P. (2015). A survey on future aspects of internet of things (IoT). *International Journal of Innovations & Advancement in Computer Science, 4*.
32. Stankovic, J. A. (2014). Research directions for the internet of things. *Internet of Things Journal, 1*(1), 3–9.
33. *How the internet of things will overcome a lack of standards*. Retrieved September 10, 2017, from https://www.cognizant.com/perspectives/how-the-internet-of-things-will-overcome-a-lack-of-standards.
34. Wang, C., Daneshmand, M., Dohler, M., Mao, X., Hu, R. Q., & Wang, H. (2013). Guest editorial - special issue on internet of things (IoT): Architecture protocols and services. *IEEE Sensors Journal, 13*(10), 3505–3510.
35. *Internet of things undermined by a lack of standards warns Pentaho VP EMEA Paul Scholey* [online]. Retrieved October 31, 2017, from http://www.computing.co.uk.
36. Newman, J. (2016). Why the internet of things might never speak a common language. [Online]. Retrieved October 31, 2017, from http://www.fastcompany.com/3057770/why-the-internet-of-things-might-never-speak-a-common-language.
37. Adjih, C., Baccelli, E., Fleury, E., Harter, G., Mitton, N., Noel, T., Pissard-Gibollet, R., Saint-Marcel, F., Schreiner, G., Vandaele, J., and Watteyne, T. (2016).
38. FIT IoT-LAB: A large scale open experimental IoT testbed. In IEEE world forum on internet of things, WF-IoT 2015 - proceedings, pages 459–464.
39. Al-Fuqaha, A., Guizani, M., Mohammadi, M., Aledhari, M., & Ayyash, M. (2015). Internet of things: A survey on enabling technologies, protocols, and applications. *IEEE Communications Surveys and Tutorials, 17*(4), 2347–2376.
40. Belli, L., Cirani, S., Davoli, L., Gorrieri, A., Mancin, M., & Picone, M. (2015). Design and deployment oriented testbed. *IEEE Computer, 48*(9), 32–40. Bloem, J. (2016).
41. Retrieved from https://www.harperdb.io/blog/iot-sensing-as-a-service.
42. Perera, C., Zaslavsky, A., Christen, P., & Georgakopoulos, D. (2013). Sensing as a service model for smart cities supported by internet of things. *Transactions on Emerging Telecommunications Technologies, 25*, 81–93.
43. Charith Perera "Sensing as a service model for smart cities supported by internet of things", Transactions on Emerging Telecommunications Technologies 2014, 25, 81.
44. Mohammadi, N. (2017). *Smart city digital twins*. IEEE Symposium Series on Computational Intelligence (SSCI).
45. Yi Ding, A., Janssen, M. (2018). *Opportunities for applications using 5G networks: Requirements, challenges, and outlook*. ICTRS'18, October 8–9, Barcelona, Spain.
46. Huh, S. (2017). *Managing IoT devices using Blockchain platform*. ICACT.
47. Nakamoto, S. (2008). *Bitcoin: Peer-to-peer electronic cash system*.

48. Dorri, A., Kanhere, S. S., Jurdak, R., Gauravaram, P. (2017). Blockchain for IoT security and privacy: the case study of a smart home. *Proceedings of the 2017 IEEE International Conference on Pervasive Computing and Communications Workshops* (PerCom Workshops), Kona, HI, USA. doi: https://doi.org/10.1109/PERCOMW.2017.7917634.
49. Samaniego, M., Deters, R. (2016). Blockchain as a service for IoT. *Proceedings of the 2016 IEEE International Conference on Internet of Things (iThings) and IEEE Green Computing and Communications (GreenCom) and IEEE Cyber, Physical and Social Computing (CPSCom) and IEEE Smart Data (SmartData)*, Chengdu, China.
50. Huh, S., Cho, S., Kim, S. (2017). Managing IoT devices using blockchain platform. In: *Proceedings of the 2017 Nineteenth International Conference on Advanced Communication Technology (ICACT)*, Bongpyeong, South Korea.
51. Fournier, G. *Challenges and solutions on architecting blockchain systems*. CASCON'18, October, 2018, Markham, ON, Canada.

IoT Physical Layer: Sensors, Actuators, Controllers and Programming

1 Introduction

The physical layer is the most detailed level of abstraction in IoT. It mainly consists of sensors that acquire information for the system and actuators that do actions in response to instructions from the system. To imagine how they both, actuators and sensors, act together in a system, a smart house is considered for example. The actuators here are used to lock and unlock doors, switch on/off the lights and alert users of any warnings or control the temperature of a room or the whole house. The sensors are used to send feedback to the controller of each small system of those systems mentioned above. For example, they send feedback about the condition of the rooms and whether there are any people in the rooms or not, and accordingly, the controller sends its signals to the actuators to turn off unnecessary working devices such as the lights and the air conditioner. Transducer terminology is used for both sensors and actuators. It means a device that converts energy form to another. In this chapter, different types of sensors and actuators are thoughtfully presented and discussed. Actuators may be written, sensors may be read. Moreover, different controllers used in IoT are discussed with its programming methods.

2 Sensors

"The Internet of Things is about empowering computers so they can see, hear and smell the world for themselves" by Kevin Ashton, the father of IoT.

One of the most essential components for IoT is the sensors. Sensors basically sense the physical phenomena or property that happens around them and sense different parameters according to the purpose of usage such as temperature, pressure, and humidity. Each sensor can only measure a unique property. Sensors are manufactured in different shapes and sizes. They can be mechanical sensors, electrical

© Springer Nature Switzerland AG 2019
K. S. Mohamed, *The Era of Internet of Things*,
https://doi.org/10.1007/978-3-030-18133-8_2

sensors, and chemical sensors. Sensors do not affect the measured property. The sensors can be classified according to their output as analog or digital sensors or according to the data type (scalar or vector). Moreover, they are divided into active and passive. Active sensors are those which require an external excitation signal or a power signal. Passive sensors, on the other hand, do not require any external power signal and directly generates output response.

One of the most **sensors requirements** are accuracy, resolution, and sensitivity. Most sensors have linear behavior. A sensor is a tiny device that measures a specific physical quantity. All IoT systems depend on the existence of one or more sensors. They are very essential in all aspects of life as they are considered a feedback to the control that gives its signal to the actuator to reach a desired goal. There are different types of sensors, including phone-based, medical, environmental, and chemical sensors. They all have light weights and single functions, in addition to being inexpensive and miniaturized devices, but constrained to the battery capacity and the ease of deployment [1, 2].

There are different types of smart phone sensors like accelerometers that sense the motion of a mobile phone, gyroscopes that detect the orientation of the mobile phone, Global Positioning System (GPS) sensors that detect the position of the mobile, light sensors, proximity sensors, magnetometers, cameras, and microphones. Accelerometers, for example, could be mechanical, using springs, cantilever beams, and seismic masses; capacitive, using capacitive plates that change the capacity with their movement; or piezoelectric, which generate electrical signals when squeezed [3].

On the other side, medical sensors are very important for healthcare applications. They can monitor very critical parameters that ease the patient's diagnosis and provide quick feedback to the doctor without the urge to go to the hospital. These parameters contain heart rate, body temperature, and blood glucose levels. Recently, there has been a new promising IoT device, called monitoring patch, which is put under the skin to monitor a certain health parameter periodically.

Neural sensors are becoming commonly used in our lives. They make it easy to infer the brain state and train it for better focus or, in other words, neurofeedback. This technology is called EEG (Electroencephalography). The communication of neurons electronically creates electric field. This electric field is measured in terms of frequency and characterized into alpha, beta, gamma, and delta waves.

Due to the very rapid changes in the environment, environmental sensors are used to measure temperature, humidity, pressure, and air and water pollution. Chemical sensors, on the other hand, detect both chemical and biochemical substances. New technologies like e-nose and e-tongue have been widely used to measure the amount of some chemicals that indicate the quality odor and taste, respectively. Examples of different sensors are summarized in Table 1. Sensors examples are shown in Fig. 1. A sensor acquires a physical parameter and converts it into a signal suitable for processing (e.g., optical, electrical, mechanical) as shown in Table 2.

Sensors can be classified according to applications as follows:

- Environmental sensing (light, pressure, temperature, humidity, etc.).
- Biometrics (fingerprinting, glucose, heart rate, breathalyzer, etc.).

Table 1 Different types of sensors and their applications: they mimic the five senses (visual, touch, smell, taste, auditory)

Five Senses	Sensor type	Application	Energy conversion	Example
Visual	IR motion sensor	Detects if there is an obstacle or not.	Measures infrared light radiating from objects in its field of view.	PIR
	Ultrasonic distance sensor	Detects how far the obstacle is from a particular point.	–	HC-SR04
	Camera sensor	Detect photos	–	–
	Speed sensor	Doppler effect sensor	–	–
	Light sensor	Light dependent resistor (LDR) Photodiode: measure light intensity	Light → electrical signals	LDR RES-0276
	Position sensor	Opto-coupler	–	–
	Tilt sensor	Detects tilt	–	–
	GPS	Location	–	–
Touch	Temperature and humidity sensor	Thermocouple Thermistor	Heat → electrical signals	DHT-11
	Pressure sensor	Detects pressure: piezometer	Pressure → electrical signals	Touchscreen sensor or piezo-resistive MEMS sensor
	Pushbutton sensor	Detects pushing a button	–	–
	Weight sensor	Detects weight	–	–
	Fingerprint sensor	Detects fingerprint	–	–
	Rain sensor	Rain drop		Rain sensor KG004
Auditory	Sound sensors	Microphone: Piezo-electrical	Sound waves → electrical signals	
Smell	Smoke detection sensor	Detects smoke	–	MQ-135
Taste	Biosensors	Detects various biological elements: electrocardiograph, sugar measure	–	–

- Communication (near-field sensors, infrared remotes, etc.).
- Mechanical sensing (gyroscopes, MEMS, etc.).

2.1 Infrared (IR) Sensor

Infrared sensor can measure temperature sensitive physical properties by the infrared ray. Infrared light has the physical properties of reflection, refraction, scattering, interference, and absorption. Anything, as long as it has a certain temperature above absolute zero, will be provided with infrared radiation. The infrared sensor measurement can be done without direct contact with the measured object directly, so there is no friction and has the advantages of high sensitivity, fast response and other advantages. The infrared sensor is composed of optical sensing system, detection element, and a switching circuit. According to different structures, the optical induction system can be divided into transmission type induction system and reflective induction system. The sensitive element is widely used in thermal resistance. Thermal resistance will be heated by infrared radiation and then the resistance will be changed. After that, the transformation of electrical signal changes response to change-over circuit. They produce and receive infrared waves in the form of heat.

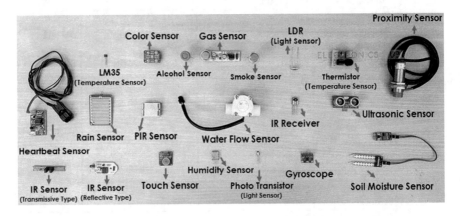

Fig. 1 Sensors examples: partial list

Table 2 Detectable phenomenon by sensors

Detectable phenomenon	Quantity
Acoustic	Wave (amplitude, phase, polarization, velocity, Spectrum)
Biological and chemical	Fluid concentrations (gas or liquid)
Electric	Charge, voltage, current, electric field (amplitude, phase, polarization), conductivity, permittivity
Magnetic	Magnetic field (amplitude, phase, polarization), flux, permeability
Optical	Refractive index, reflectivity, absorption
Thermal	Temperature, flux, specific heat, thermal conductivity
Mechanical	Position, velocity, acceleration, force, strain, stress, pressure, torque

2.2   Temperature/Humidity Sensor

Most physical, electronic, chemical, mechanical, and biological systems are affected by temperature. There are many types of temperature sensors such as thermocouple sensors and thermistors. A thermocouple is a device consisting of two different and dissimilar conductors in contact. It produces a voltage as a result of the thermoelectric effect. Thermocouple sensor is made by joining two dissimilar metals at one end. The thermistor is a temperature sensing device whose resistance changes with temperature. Thermistors, however, are made from semiconductor materials. Humidity sensors use capacitive measurement by relying on electrical capacitance.

2.3   Pressure Sensor

Pressure sensors are used to measure the pressure of gases or liquids including water level, flow, speed, and altitude. Practical examples include sensors for pumps and compressors, hydraulic systems, and refrigerators. A pressure sensor typically acts as a transducer where it generates a signal as a function of the pressure imposed. Touch screen smartphones, tablets, and computers come with various pressure sensors. Whenever slight pressure is applied on the touch screen through a finger, tiny pressure sensors determine where exactly pressure is applied and consequently generate an output signal that informs the processor.

A MEMS air pressure sensor works on two principles. The first is called the piezoelectric effect. Piezoelectric materials generate an electric current. When they're subject to force, they're deformed from their original shape, which allows for two different equations. The electric current can be used to calculate the deformation of the material, and the deformation of the material can be used to calculate the force, and thus the current can be used to calculate the force. These calculations can be performed in reverse order as well.

The second principle is that a stationary system has a net force of zero. Think of an inflated balloon. The size of the balloon is determined when the contracting force of the material equals the difference in pressure inside and outside of the balloon. Pressure is a force applied over an area so, in other words, the forces around the balloon are equal. You can force the balloon to be smaller by pushing it under water because there will be a larger outside pressure underwater. If you climbed a mountain, the balloon would get larger because the outside pressure would be less [4].

2.4   Global Position System (GPS)

GPS is composed of a space satellite, a ground signal connecting point, and a user signal receiving device. It can provide users with high precision position, speed and temporal information in all weather real time. Compared with the positioning function, it's more important and widespread for GPS to apply in power system. The monitoring and protection system in electric power system such as microcomputer protection and security automatic equipment monitoring system, dispatching automation system, wave recorder automation equipment fault accident, all need accurate time standard to achieve accurate synchronization purposes.

2.5 Proximity Sensor

A proximity sensor creates a net of electric/magnetic field and detects an object which enters the field, just as a spider forms its web and catches its prey. The net is created by the magnetic lines originated from the oscillation circuit. When a metallic object comes into the field, the magnetic lines get disordered, which is transmitted to the oscillating circuit. The oscillating circuit will detect the object approaching and output the decision. Si114x and Si1102 are typical examples of proximity sensors used in IoT.

2.6 Image Sensor

Image sensors are an important type of sensor in several emerging Internet of Things (IoT) applications. They provide detailed environment information by a large array of photodiodes and have a highly competitive market price due to their standardization.

2.7 Smart Passive Sensors

Smart Passive Sensors (SPS devices) are battery-less, microcontroller-less RF sensor nodes that measure moisture or temperature, and can be manufactured in PCB, flexible PET, or foam-based form factors. Without a battery, SPSSensors SPS tags use existing RF field to generate 30 dBm of power that it **harvests** for operation. Current-generation SPS tags include a memory block of a few hundred bytes for storing sensor values, and wireless transmit range of 1–2 m to 5–10 m depending on SKU.

2.8 Ultrasonic Sensor

An ultrasonic sensor is a non-contact type device that can be used to measure distance as well as velocity of an object. An ultrasonic sensor works based on the properties of the sound waves with frequency greater than that of the human audible range.

2.9   Accelerometer

It measures acceleration in one or more directions, and position can be deduced by integration. It uses mass spring method and measures the capacitance to create output. Moreover, there is 3D accelerometer to measure accelerations in three directions.

2.10   Gyroscopes

It measures rotational angles. Modern implementations are using Micro-Electro-Mechanical Systems (MEMS) technologies. It can be used for self-balancing robot. Gyroscope is accurate in high frequency measurement while accelerometer is accurate in low frequency measurement. So, we can combine two sensors to find output at all frequencies. The difference between accelerometer and the gyroscope is accelerometer measures linear acceleration based on vibration, whereas the gyroscope is intended to determine an angular position based on the principle of the rigidity of space.

2.11 CO_2 Gas Sensor

CO_2 sensor measures gaseous CO_2 levels in an environment and measures CO_2 levels in the range of 0–5000 ppm. Moreover, it monitors how much infrared radiation is absorbed by CO_2 molecules.

2.12   Solar Cell Sensor

Photovoltaics are best known as a method for generating electric power by using solar cells to convert energy from the sun into a flow of electrons by the photovoltaic effect [5].

2.13   LiDAR Sensor

The LiDAR instrument emits rapid laser signals, sometimes up to 150,000 pulses per second. The signals bounce back from the obstacles. The sensor positioned on the instrument measures the amount of time it takes for each pulse to bounce back. Thus, the instrument can calculate the distance between itself and the obstacle with accuracy. It can also detect the exact size of the object. LiDAR is commonly used to make high-resolution maps [6].

2.14 RADAR Sensor

The RADAR system works in much the same way as the LiDAR, with the only difference being that it uses radio waves instead of laser. In the RADAR instrument, the antenna doubles up as a radar receiver as well as a transmitter. However, radio waves have less absorption compared to the light waves when contacting objects. Thus, they can work over a relatively long distance. The most well-known use of RADAR technology is for military purposes. Airplanes and battleships are often equipped with RADAR to measure altitude and detect other transport devices and objects in the vicinity [6].

2.15 Optical Sensors

The optical sensors convert light rays into an electronic signal; it measures a physical quantity of light and transforms into a form which is readable, maybe digital form. It detects the electromagnetic energy and sends the results to the units. It involves no optical fibers. It is a great boon to the cameras in mobile phones. Also, it is used in mining, chemical factories, refineries, etc. LASER and LED are the two different types of light source. Optical sensors are integral parts of many common devices, including computers, copy machines (Xerox), and light fixtures that turn on automatically in the dark [7].

3 Actuators

One of the most essential components for IoT is the actuators. Actuators are basically performing some actions based on the readings of the sensors and the required specifications which differ from an application to another. An actuator requires a control signal and a source of energy. There are three different types of actuators: mechanical, electrical, and pressure. Actuators convert energy to motion. An actuator is a device that converts an electrical signal into a mechanical signal or any other useful form of energy. Some examples include speakers, heaters, cooling elements, and displays. They can be electrical, hydraulic, or pneumatic actuators depending on their theory of operation. For example, hydraulic actuators use fluid mechanics to facilitate motion, whereas pneumatic actuators make use of the compressed air to generate pressure difference. Examples of different actuators are summarized in Table 3 [8].

3.1 Electrical Actuators

Electric actuators are devices driven by small motors that convert energy to mechanical torque. The created torque is used to control certain equipment. Actuators are also used in engines to control different valves.

3.2 Mechanical Actuators

Mechanical actuators convert rotary motion to linear motion. Devices such as screws and chains are utilized in this conversion. The simplest example of mechanical liner actuators is referred to as the "screw," where leadscrew, screw jack, ball screw, and roller screw actuators all operate on the same principle: By rotating the actuator's nut, the screw shaft moves in a line.

3.3 Hydraulic Actuators

Hydraulic actuators are simple devices with mechanical parts that are used on linear or quarter-turn valves. They are designed based on Pascal's law: When there is an increase in pressure at any point in a confined incompressible fluid, then there is an equal increase at every point in the container. Hydraulic actuators comprise a cylinder or fluid motor that utilizes hydraulic power to enable a mechanical process. The mechanical motion gives an output in terms of linear, rotary, or oscillatory motion. **Propulsion thrusters** are an example of hydraulic actuators.

Table 3 Different types of actuators and their applications

Actuator type	Application	Energy conversion
Sound actuator	Loudspeaker	Electrical signal- > sound waves
Relay	Electromechanical switch	–
Valve	Control flow of liquid	–
Mechanical	Gears	Rotary motion- > linear motion
Thermal	Heater	–
Magnetic	Generate forces which impact on the motion of a part in the actuator	–
Electrical	Motor	Electrical signal- > motion
Hydraulic	Industrial process control	Mechanical- > linear, rotary motion
Pneumatic	Automation control	Pressure - > force

3.4   Pneumatic Actuators

Pneumatic actuators work on the same concept as hydraulic actuators except compressed gas are used instead of liquid.

4   IoT Hardware Platforms

The controller is the device that receives the sensors' signals, processes them and makes computations on them, and then sends instruction signals to the actuators. Usually in control systems, these instruction signals are based on the difference between the sensors' readings and the desired values of the physical quantities, and thus these instruction signals are sent to the actuators in order to set the system back to the desired physical quantities' values. There are many hardware platforms with different capabilities that can be used in IoT applications. Choosing which hardware platform is used is based on the requirement of the IoT applications and depends also on whether we need it for development only or mass production. The factors that define the hardware platform for IoT applications are [9–14]:

- Reduction of employed transistors: This will reflect on the die size, packaging, and unit cost. The progress made on transistor area decreases the cost, but leakage power dominates on the overall chip.
- Time-to-market: It is the main factor that guides the design to the proper platform. The market requires a generic solution to apply its demands, therefore time is a very critical factor to choose what type of platform is needed, which might be in most cases an expensive one.
- Nonrecurring Engineering (NRE) costs: It is the cost of the development process for the IoT platform, either software or hardware. This does not only imply sustainability in reliable systems, but also the ability to develop the platform in less time as much as possible.

Based on the previous factors, the designer is capable of choosing the perfect platform out of the following:

- Application-Specific Integrated Circuit (**ASIC**): ASIC is a well-established process designed, as the name suggests, for a specific application. The fabricated ASIC chips give very optimal performance with the lowest number of transistors, and most importantly, the least power consumption. In addition, the technology is very cheap when mass produced. However, ASIC is not usually used, because it consumes time and resources to develop. In brief, it has large time-to-market, which makes the industry seek other faster generic solutions.
- Field Programmable Gate Array (**FPGA**): FPGAs provide a more generic solution that is required in industry. They have less time-to-market and NRE costs to develop their products. However, they consume much more power than ASIC chips, which is one of the most challenging issues in IoT. In addition, they are

very expensive, so they are used in applications with minimum number of units needed [15].

- **Microprocessors**: Microprocessors are used as a platform for building IoT devices. Some chips have the microprocessor together with other blocks such as RAMs and different other modules. In this technique, the whole system is built on the chip with all its peripherals. The system acts as a gateway for the local devices to the Internet. This requires that the chip must support several protocols to facilitate the communication between local devices and sensors with the microcontroller (Bluetooth, ZigBee) as well as sending and receiving data from the cloud (Wi-Fi, Ethernet). There are several systems used commercially for this purpose. Arduino family (Uno, Yun) is based on ATmega32U4 processor with its peripherals to do the complete functions. Other chips are used such as Raspberry Pi, which uses Broadcom BCM283 (5~7) SoC. These platforms are generic and can be used for several applications. As a result, hardware overheads are installed. This increases the power consumption for the system. Therefore, more optimization is required to save the battery for the longest time.
- **Open-source embedded systems**: Arduino and Raspberry Pi, Intel Galileo.

4.1   Arduino: Atmel-Based

Arduino is very popular in IoT applications as it is very cheap and easy to use and to build a fast prototyping. It is an open-source hardware and software (Arduino IDE). The architecture of Arduino UNO is shown in Fig. **2**. The main features of Arduino UNO are summarized in Table 4. There are many other families of Arduino such as Mega with more input and output pins. The programming of Arduino is based on C and C++. We may need memory extension to Arduino to support more services. Shields aren't the only way to extend an Arduino board—you can hook sensors to it. These are some of the hundreds (if not thousands) available. Many of these are not made specifically for Arduino. Listing **1** shows a pseudo-code for turning an LED on and off.

4.2   Raspberry Pi: ARM-Based

The Raspberry Pi is a very small computer as depicted in Fig. 3. The programming of Raspberry Pi is based on C, C++, and Python. Compared to Arduino, it is more powerful in terms of processing, memory, and features. So, it is useful in multimedia applications which need more resources. But, it comes with more cost. The main features of Raspberry Pi 3 are summarized in Table 5.

The Raspberry Pi has the ability to interact with the outside world, and has been used in a wide array of digital maker projects, from music machines and parent detectors to weather stations and tweeting birdhouses with infrared cameras. The following operating systems can run on Raspberry Pi:

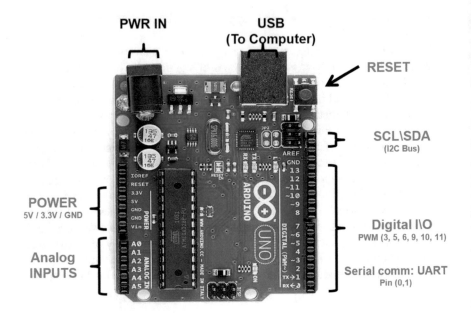

Fig. 2 Arduino development kit [16]

Table 4 Arduino UNO features

Feature	Value
Operating voltage	5 V
Operating frequency	16 MHz
Digital I/O pins	14
Analog input pins	6
PWM	6
UART	1
I2C	1
USB	1

- **Raspbian**: A free Debian-based OS optimized for Raspberry Pi's hardware. Raspbian comes with all the basic programs and utilities you expect from a general-purpose operating system. Supported officially by the Raspberry foundation, this OS is popular for its fast performance and its more than 35,000 packages. The easiest way to install Raspbian on your Pi is by deploying its image file onto an SD card.
- **Ubuntu MATE**: Ubuntu MATE is a stable and simple OS, which brings a configurable yet still light-on-resources MATE desktop for its users. It is especially good for devices short on hardware specs, making it perfect for Raspberry Pi devices that can't run a composite desktop.
- MATE desktop comes with essential apps like a file manager, text editor, image viewer, system monitor, document viewer and terminal.

Listing 1 Pseudo-code for turning an LED on and off

```
int ledPin =  13;

void setup()  {

  pinMode(ledPin, OUTPUT);

}

loop()

{

  digitalWrite(ledPin, HIGH);

  delay(500);

  digitalWrite(ledPin, LOW);

  delay(500);

}
```

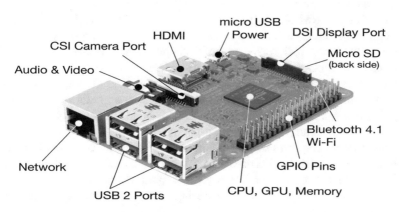

Fig. 3 Raspberry Pi 3 development kit [17]

- **Snappy Ubuntu**: A lightweight edition of the popular Ubuntu OS aimed for clouds and devices. Snappy Ubuntu Core uses a minimal server image with the same system libraries. Applications run noticeably faster and are more reliable and secure because of the transactional systems management (like Docker), hence the term "Snappy."
- **Pidora**: Pidora is a remix of the well-known Fedora operating system for Raspberry Pi. Designed from the latest build of Fedora for the ARMv6 architecture, Pidora allows greater speed and carries applications and components from the Fedora 20 package set.
- **Linutop**: An OS that can be quickly set up on a Raspberry Pi. Linutop uses a Raspbian-base with classic and lightweight XFCE graphical environment. It's also handy for secure professional uses, such as in kiosks casting public access or in embedded systems like electronic devices.

Table 5 Raspberry Pi 3
features

Feature	Value
Operating voltage	5 V
Operating frequency	1.2 GHz
GPIO	40
HDMI	1
Ethernet	1
GPU	1 @ 400 MHz
Bluetooth 4.0	1
SDCARD (min 2 GB)	1
CSI camera port	1
WIFI	1

- **SARPi**: Short for "Slackware ARM on a Raspberry Pi." SARPi is a community product of Slackware Linux enthusiasts. Considered widely as one of the best OS choice for Raspberry Pi, this can be installed on an 8 GB SD card. Although the ARM version doesn't support all the apps, but most applications (including essential ones) have been ported for the ARM architecture.
- **Arch Linux ARM**: A version of Arch Linux ported for ARM computers. Arch Linux ARM offers versions 6 and 7 for Raspberry Pi and Raspberry Pi 2, respectively. Its design philosophy promotes simplicity and user-centrism, ensuring that Linux users are in full control of the system.
- **Gentoo Linux**: An open-source Linux-based computer OS. Gentoo Linux compiles source code locally according to the user's preferences to uphold performance. For this reason, Gentoo Linux's builds are often optimized for a specific type of computer, such as Raspberry Pi.
- **FreeBSD**: A computer OS used to power servers, embedded systems as well as computers. FreeBSD offers advanced networking, security, and storage features. Its powerful networking services make it the platform of choice when setting up an Internet or Intranet server, thus ensuring fast response times and robust memory management.
- **Kali Linux**: Kali Linux is an advanced penetration platform with versions designed to support Raspberry Pi. A Debian-based Linux distribution, this OS has several tools for information security operations such as penetration testing, forensics, and reverse engineering. It's not limited to those operations as it is suitable for a general-purpose OS too.
- **RISC OS Pi**: RISC OS Pi is the latest version of the RISC OS designed for Raspberry Pi. RISC OS Pi brings an alternative desktop environment and a stack of heavily functional applications for the Pi board. If creating a boot image is too much work, you can get a specially prepared SD card preloaded with the RISC OS.

Moreover, Raspberry pi supported the following programming languages:

1. SCRATCH
2. PYTHON
3. HTML5
4. JAVASCRIPT
5. JQUERY
6. JAVA
7. C PROGRAMMING LANGUAGE
8. C++
9. PERL
10. ERLANG

Raspberry Pi had released several generations:

- Raspberry Pi Model B (First generation)—February 2012
- Raspberry Pi Model A—February 2013
- Raspberry Pi Compute Model—April 2014
- Raspberry Pi Model B+—July 2014
- Raspberry Pi 2—February 2015
- Raspberry Pi Zero—November 2015
- Raspberry Pi 3 Model B—February 2016
- Raspberry Pi Zero W—February 2017
- Raspberry Pi 3 Model B+—March 2018

4.3   Intel Galileo

Intel Galileo combines Intel technology with support for Arduino ready-made hardware expansion cards (called "shields") and the Arduino software development environment and libraries. The development board runs an open-source Linux operating system with the Arduino software libraries, enabling reuse of existing software, called "sketches." The sketch runs every time the board is powered. Intel Galileo can be programmed through OS X, Microsoft Windows and Linux host operating software. The board is also designed to be hardware and software compatible with the Arduino shield ecosystem. Figure 4 shows Intel Galileo Gen. 2.

4.4   Tessel

Tessel is a microcontroller that runs JavaScript. It is a development board with onboard WiFi capabilities that allows you to build scripts. Figure 5 shows Tessel development kit. The platform is built using high-performance 88MW32x Cortex-M4F Microcontroller and low-power 802.11 b/g/n Wi-Fi SoCs.

4.5 AVR-IoT

Microchip and Google have partnered to provide you with the ideal foundation for building your next cloud-connected design. Combining a powerful AVR® microcontroller, a CryptoAuthentication secure element IC and a fully certified Wi-Fi network controller these boards offer the most simple and effective way to connect embedded applications to Google's Cloud IoT core platform. Figure 6 shows AVR-IoT development kit.

4.6 Marvell

The Marvell® Wi-Fi microcontroller platform provides a highly cost-effective, flexible, and easy-to-use hardware/software platform to build a new generation of smart connected devices delivering a broad range of services to consumers including thermostats, appliances, lighting, home automation, and remote access. Figure 7 shows the development kit for Marvell IoT hardware platform.

4.7 ARM

With the challenge of security dominating IoT application development, ARM®'s latest v8-M microcontrollers have been designed to reduce the complexity of developing secure embedded solutions, whether they are for small devices or complex SoCs. Figure 8 shows the development kit for ARM IoT hardware platform.

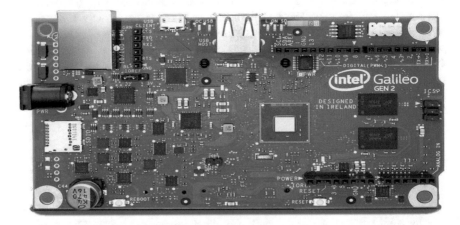

Fig. 4 Intel Galileo Gen. 2 development kit [18]

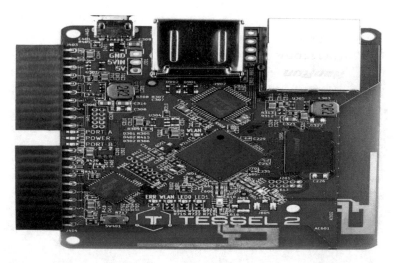

Fig. 5 Tessel development kit [19]

Fig. 6 AVR-IoT development kit [20]

4.8 Particle Electron

It uses the STM32F205 microcontroller. It presents 36 total pins, such as UART, SPI, I2C, and CAN bus. Electron provides 1 MB of Flash and 128 k of RAM. If we compare Electron with Arduino, the first one is a competent board. The hardware design for the Electron is open source. It includes a SIM card, with a global cellular network for connectivity in 100+ countries, and cloud services. All Electron family products can be set up in minutes using the Particle mobile app or browser-based setup tools [23]. Figure 9 shows the photon model IoT hardware platform.

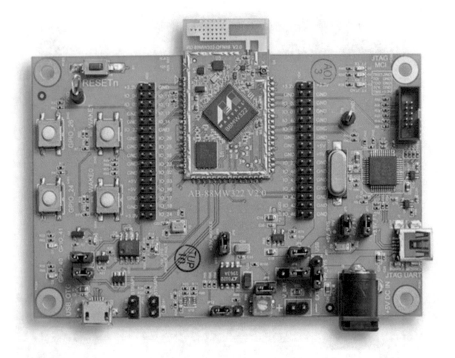

Fig. 7 Marvell IoT hardware platform [21]

4.9 NodeMCU Dev Kit

The NodeMCU is an open-source, single-board microcontroller, and low-cost, simple and smart IoT development board with a few simple Lua scripts. It gives high-level interface to hardware with simple configurations. Based on the Lexin esp8266 NodeMCU development board, with GPIO, PWM, I2C, 1-Wire, ADC and other functions, combined with NodeMCU firmware to provide the fastest way for your prototyping. It can be powered by USB, with a memory of 128KBytes and a storage of 4 MB. Figure 10 shows NodeMCU development kit.

5 IoT Software and Programming

In order for the hardware to perform well, operating systems should be installed. Operating systems organize the usage of hardware. For IoT applications, low-power and small hardware overhead operating systems should be used.

The software platform is necessary to recognize the received data, identify the needed manipulation for the desired action by the user, and transmit efficiently the new data to the right node.

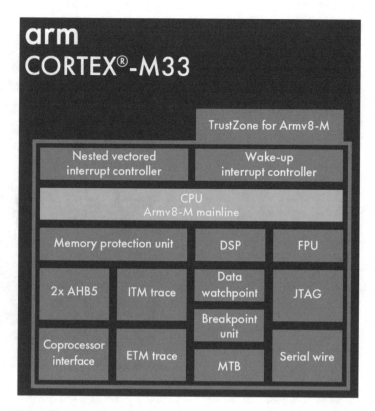

Fig. 8 ARM IoT hardware platform [22]

There are multiple of business operating systems such as IBM Watson platform as well as open-source platforms such as Linux and RIOT. Choosing the right operating system is a crucial move in order to build the optimal IoT system for the desired application. In this section, the key parameters to choose the suitable operating system (OS) are investigated as follows [25].

- IoT heterogeneous hardware support: A lot of IoT systems usually work on different types of hardware from 16-bit microcontrollers to FPGAs based on the implemented hardware. Therefore, the operating system shall be compatible with the implanted hardware platform in order to achieve excellent performance.
- Real-time operating systems: One of the most important factors that guide most of IoT designs is that whether the operating system supports predictability or not. Predictability allows the system to be in an alert position based on earlier received data. This helps the software take actions rapidly, especially in situations like fires and accidents on the road. In turn, this is great evidence on how predictability reflects the degree of smartness of IoT systems, which is in high need for development.

Fig. 9 Photon IoT
hardware platform

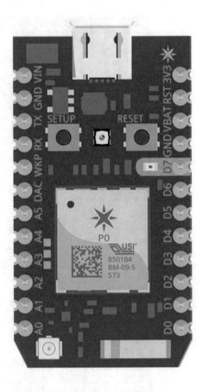

Fig. 10 NodeMCU Dev Kit [24]

- Developer friendly: IoT evolves rapidly, which requires an easy platform to apply new solutions and add smart features to the desired applications. In order to have low time-to-market, it is required to decide what operating systems are sufficient for developing the desired IoT system. Working on a developer friendly platform is highly required for companies in order not to be delayed for the market, which has its drawbacks.
- Memory: The operating system shall have low size in order to fit in the internal memories of microcontrollers. On the other hand, extra memory is needed to fulfill the size of the operating system.
- Security: Security is a very important factor that should be considered in all the layers of IoT. In order to achieve a good level of security, the operating system should support encryption engines, secure boot functions, and usage of wireless authentication protocols.
- Accommodation for low power: Despite the fact that the hardware platform takes power issues into account, operating systems that allow power management capabilities gain much focus from IoT designers. Power management feature helps increase the battery life, especially for end-nodes. For example, the software can operate only with the required area of the hardware for simple functions as long as there is no necessity to perform repeated calculations, which is the case in most times. Focusing on such a thing increases the reliability of IoT products.
- Support required for communication and network: IoT requires different types of protocols in small, medium, and large signals. The software has to strongly support such integration.

One of the operating systems that are designed especially for IoT is RIOT. The smart operating system tries to save the most possible power and area for the smart device. RIOT is an open-source operating system which requires small RAM and ROM to boot. The architecture of RIOT is microkernel based, where the user application can address different layers in the operating system. One of the important advantages of RIOT-OS is that it supports C/C++ programming languages, allowing wide range of programmers to make the best use of the OS to serve the desired application with low power as long as it can handle heterogeneous hardware projects. RIOT-OS facilitates the communication with the sensors and actuators as it supports different communication protocols. Furthermore, RIOT supports multi-threading, increasing parallelism. Many IoT applications require real-time operations; that is why RIOT-OS enforces regular kernel periods to divide the tasks and work on all of them in real time [26–28].

This guarantees a promising future for RIOT-OS in the IoT industry, especially that it requires small amount of memory and power. In addition, its programmability allows it to serve a wide range of applications. On the other hand, it might not be easy to port RIOT to developer boards and different microprocessors.

The most used operating system nowadays in IoT platforms is Linux. Linux is an open-source operating system that allows developers to modify it to give the best response. It supports C/C++ programming languages as well as multi-thread technology. It is now the most established and used software platform in IoT designs. However, it requires relatively large memory (RAM and ROM) to operate. The hardware overhead in memory as well as the more required power to operate is not ideal for real-time processing and thus hinder Linux implementation in various IoT devices. Some other business software tools are used for IoT projects such as IBM Watson and Amazon Web Services (AWS). IBM Watson is one of the leading software tools for IoT devices. It was released in 2014 and modified after that to include augmented reality, cognitive capabilities, and more capabilities to spread the operating system to many devices and developers. However, developers found several difficulties to develop the operating system to do the required functions. AWS is one of the best performing operating systems in IoT fields as it has its own cloud service as well as a user-friendly interface. Embedded C, Python (is a general-purpose programming language), and JavaScript can be used to program most of IoT hardware platforms.

5.1   Python

Python is a general-purpose programming language like C, Java, etc. but is higher level. Python features can be summarized in the following points [29]:

1. Less line of codes compared to C or JAVA.
2. Vast function library geared towards strings and files.
3. Save coding time and automate a lot of tasks.
4. Intuitive, code is concise, but human readable.

5.1.1   Python Versus Java

Although performance is not always a problem in software, it should always be a consideration. Where network I/O costs or database access dominate, the specific efficiency of a language is less significant than other aspects of technology choice and design when it comes to overall efficiency [30]. Although neither Java nor Python is especially suited to high-performance computing, when performance matters, Java has the edge by platform and by design. Although some python implementations. A lot of Java efficiency comes from optimizations to virtual machine execution. A JVM can translate bytecode into native machine code as a program executes. This Just-In-Time (JIT) compilation is why Java's performance can often rival that of native languages. Relying on JIT is a reasonably portable assumption as HotSpot, the default Oracle JVM, offers it.

Java has had support for concurrency from its first public version, whereas Python is more resolutely a sequential language. This has implications for taking advantage of current multi-core processor trends, with Java code more readily able to do so. Both Java and Python enjoy a seemingly endless supply of open-source libraries populated by code from individuals and companies who have solved common and uncommon problems, and who are happy to share so others can take advantage of their solutions. Indeed, both languages have benefited from—and been shaped by—online forums and open-source development. Though Java's been in the mainstream longer than Python (even though both are about the same age—Python being slightly older). Thus it's probable that the libraries/tools for Java are more mature and/or capable. But that's very debatable [31].

5.1.2 Python Versus C++

C++ is a general-purpose programming language. It is also developed from the original C programming language. C++ is a statically typed, free-form, multi-paradigm and a compiled programming language. Python is another programming language. However, it is quite different than C++. Python is a general-purpose, high-level programming language. Python is considered to be cleaner and more direct, with emphasis code readability [32].

C++ is now commonly used for hardware design. The design is first described in C++. It is then analyzed, architecturally constrained, and scheduled to create a register-transfer level hardware description language. It would do this through high-level synthesis.

An advantage of Python is that its code is quite shorter than most other programming languages. This allows programmers to express concepts in fewer lines of code than in C or C++. Python's language provides constructs. These constructs are intended to enable clear programs on both a small and large scale. Another advantage of Python is that it multiple programming paradigms, including object-oriented, imperative, and functional programming styles. It features a dynamic type system and automatic memory management. It also has a large and comprehensive standard library. All of which helps improve Python's usability. Also, python interpreters are available for many operating systems.

Python is a dynamic language and like other dynamic languages, it is often used as a scripting language. However, it is also often used in non-scripting contexts. Furthermore, Python code can be packaged into standalone executable programs by using third-party tools. Difference between Python and C++ can be summarized in the below points:

- Python uses garbage collection whereas C++ does not.
- C++ is a statically typed language, while Python is a dynamically typed language.
- Python is easier to use than C++.
- Python is run through an interpreter while C++ is pre-compiled.
- Hence, C++ is faster than Python.
- C++ supports pointers and incredible memory management.
- Python supports very fast development and rapid, continuous language development.
- Python has less backwards compatibility.

- Majority of all applications are built from C++.
- Majority of all 3D applications offer Python access to their APIs.
- Python code tends to be 5–10 times shorter than that written in C++.
- In Python, there is no need to declare types explicitly.
- Smaller code size in Python leads to "rapid prototyping," which offers speed of development.
- Python requires an engine to run.
- Python is interpreted each time it runs.
- Python is hard to install on a Windows box and thus makes distribution of the program problematic.
- C++ is a pure binary that links to existing libraries to assist the coding.
- In Python, variables are in scope even outside the loops in which they are first instantiated.
- In Python, a function may accept an argument of any type, and return a value of any type, without any kind of declaration beforehand.
- Python provides flexibility in calling functions and returning values.
- Python looks cleaner, is object oriented, and still maintains a little strictness about types.

5.2   JavaScript

JavaScript has proven itself worthy on both the client and server side world of web applications; it has potential in the ever expanding Internet of Things (IoT) as much of the internet already speaks JavaScript. JavaScript has a range of existing libraries, plugins, and APIs, many of which can be utilized in the Internet of Things. JavaScript is quite good at event driven applications. These are the sorts of applications in which each device listens for various events and responds when events occur that it cares about. JavaScript has matured as a language [33].

5.3   C/Embedded C

If you use C on computers, then it is called regular C. If you use C on electronics, then it is called "Embedded C." C is also a common language for microcontroller programming, making it a no-brainer for sensor and gateway hardware layer applications. However, since C is such a low-level language, its syntax can become cluttered and messy quickly if developers aren't fluent in best practices. At the device level, computing power is usually quite limited. C works best here because the language is ideal for writing low-level code (i.e., code close to the hardware layer), it doesn't require a lot of processing power, and it's able to work directly with the RAM. Each processor has its own C library and its own compiler. But, it is the same concepts. Embedded C includes extra features over C. The main difference between C on computers and C on electronics is input/output [34].

5.4   R Language

R is an open-source programming language meant for conducting statistical calculations on datasets. The analysis of those sets is used to create data structures to be graphed for data visualization. R in its early days was developed with terminals in mind. That meant having to type code line by line to create the data visualization. Various graphic user interfaces have been developed over time. The most popular among professionals is R Studio [35].

5.5   Swift

Swift is the programming language that is used for creating the apps for MacOS or Apple's iOS devices. If you want to interact with the iPhones and iPads with your central home hub, swift is the way. Swift is gaining more fame as a programming language that its processor Objective-C. Apple to achieve its goal of becoming a leader of IoT at home, is building libraries. These libraries can handle much of the work, it will make be easier for developers to focus on the task. While the HomeKit platform handles the integration [35].

5.6   PHP

PHP is being added by the developers to their pile of codes. The code's main objective is to juggle micro services on the server. They can turn the lowliest thing of the internet into a full web server. With the help of PHP, apps are developed using the GPS data from IoT devices. Therefore, it is difficult choosing any one of the language because all are the best. C, Python, and Java are the most popular IoT programming languages [36].

6   Conclusions

This chapter discusses the IoT physical layer which includes sensors, actuators, and controllers. Moreover, it discusses the most important languages used in programming IoT hardware platforms. Variety of IoT hardware platforms are introduced that can support entire development of IoT applications and systems. Moreover, different IoT programming languages are discussed.

References

1. Ramachandran, B. (2017). *IoE/IoT | anything connected*. Connectedtechnbiz.wordpress.com. [Online]. Retrieved July 4, 2017, from https://connectedtechnbiz.wordpress.com/category/ioeiot/
2. Sethi, P., & Sarangi, S. (2017). Internet of things: Architectures, protocols, and applications. *Journal of Electrical and Computer Engineering, 2017*, 1–25.
3. Retrieved from http://www.electrical4u.com/sensor-types-ofsensor/
4. Retrieved from https://www.iotforall.com/mems-2-iot-pressure-sensors/
5. Chong, K., Khlyabich, P. P., Hong, K., Reyes-Martinez, M., Rand, B. P., & Loo, Y. (2016). Comprehensive method for analyzing the power conversion efficiency of organic solar cells under different spectral irradiances considering both photonic and electrical characteristics. *Applied Energy, 180*, 516–523.
6. Retrieved from https://www.sensorsmag.com/components/lidar-vs-radar?fbclid=IwAR0kffF H1cmcuXXfGu7MG5kZS3EvKqVGMCslA8VFqHJ2nYzSSZmRQwjAgNs
7. Retrieved from https://iot4beginners.com/commonly-used-sensors-in-the-internet-of-things-iot-devices-and-their-application/
8. Retrieved from http://www.thegreenbook.com/four-types-of-actuators.
9. Perera, C., et al. (2014). A survey on internet of things from industrial market perspective. *IEEE Access, 2*, 1660–1679.
10. Al-Fuqaha, A., et al. (2015). Internet of things: A survey on enabling technologies protocols and applications. *IEEE Communication Surveys and Tutorials, 17*(4), 2347–2376., Fourth quarter.
11. Products | IoT Solutions – ARM. ARM | The Architecture for the Digital World, 2016. [Online]. Retrieved from https://www.arm.com/products/iot-solutions. [Sep. 24 2016].
12. IoT hardware guidebook | 2016 prototyping boards and development kits. Postscapes. com, 2016. [Online]. Retrieved September 24, 2016, from http://www.postscapes.com/internet-of-things-hardware/.
13. IoT Platform | Simplify Data Device and Embedded Apps Management. Eurotech.com, 2016. [Online]. Retrieved September 24, 2016, from https://www.eurotech.com/en/products/iot+platform.
14. IoT Security and Scalability on Intel® IoT Platform. Intel, 2016. [Online]. Retrieved September 24, 2016, from http://www.intel.com/content/www/us/en/internet-of-things/iot-platform.html.
15. Mohamed, K. S. (2016). *IP cores design from specifications to production: Modeling, verification, optimization, and protection. IP cores design from specifications to production* (pp. 13–50). Cham: Springer.
16. Retrieved from https://www.arduino.cc/
17. Retrieved from https://www.raspberrypi.org/
18. Retrieved from https://www.arduino.cc/en/ArduinoCertified/IntelGalileo
19. Retrieved from https://tessel.io/
20. Retrieved from https://www.avr-iot.com/
21. Retrieved from https://www.marvell.com/microcontrollers/wi-fi-microcontroller-platform/
22. Retrieved from https://community.arm.com/processors/trustzone-for-armv8-m/b/blog/posts/nordic-announce-first-cortex-m33-based-chip-with-trustzone?_ga=2.232749215.1070669737.1544091187-1453952905.1543909720.
23. Particle Community. Retrieved October 15, 2018, from https://community.particle.io/
24. Retrieved from https://en.wikipedia.org/wiki/NodeMCU
25. Slant. (2018). *What are the best programming languages for IoT (internet of things)?* Accessed 10 June 2018.
26. IoT Operating Systems, Arrow, 2016. [Online]. Retrieved July 4, 2017, from https://ww.arrow.com/en/research-and-events/articles/iot-operating-systems.
27. Baccelli, E., Hahm, O., Gunes, M., Wahlisch, M., Schmidt, T. C. (2013). *RIOT OS: Towards an OS for the internet of things*. 32nd IEEE International Conference on Computer Communications (INFOCOM), 2013.

28. Pelino, M., Hewitt, A. (2016). *The FORRESTER wave™: IoT software platforms, Q4 2016.* FORRESTER, 2016.[online]. Retrieved July 23, 2017, from https://kloudrydermcaasicmfor-rester.s3.amazonaws.com/mcaas/Reprints/RES136087.pdf . [Accessed: 23- Jul- 2017].
29. Vrat, S. (2017). *IoT with Python: Essential packages.* Accessed 10 June 2018.
30. Henney, K. (2017, March 21). *Java vs. Python: Which one is best for you?* Retrieved from blog.appdynamics.com: https://blog.appdynamics.com/engineering/java-vs-python-which-one-is-best-for-you/.
31. Delgado, R. (2017). *Why Java is the language of choice for the internet of things (IoT).* KD Nuggets. Accessed 10 June 2018.
32. Retrieved from www.lmpt.univ-tours.fr/~volkov/C++.pdf.
33. Retrieved from https://www.tutorialspoint.com/javascript/javascript_tutorial.pdf.
34. Retrieved from www.kciti.edu/wp-content/uploads/2017/07/cprogramming_tutorial.pdf.
35. Retrieved from https://www.dmnews.com/data/news/13055403/microsoft-adding-an-r-to-iot
36. Retrieved from https://techbeacon.com/app-dev-testing/top-6-programming-languages-iot-projects

IoT Networking and Communication Layer

1 Introduction

The communication layer is considered as the backbone of the IoT systems. It is the main channel between the application layer and different operating activities in the IoT system. The whole physical system is loaded with amounts of data and information that need to be shared with other nodes. Therefore, it is needed to set up a suitable connection network among these nodes through a communication protocol. The communication could be wire-connected or wireless based on the protocol defined by the designer. Moreover, networks are very vital components in IoT to connect things to the outside world of internet. IoT requires an intelligent network infrastructure. Any IoT hardware can connect to the internet via the following [1–17]:

- **Ethernet** (built-in or shield)
- **Wi-Fi** (module)
- **5G** (module)
- **Bluetooth/BLE** (module), via 4G/Wi-Fi of phone
- **ZigBee** (module), via ZigBee gateway
- **USB** (built-in), via desktop computer
- **RFID**
- **Satellites**

Communication media classification according to connectivity for IoT is shown in Fig. 1. The requirements of IoT communications include: energy efficiency, range, cost, reliability, security, delay, and scalability.

© Springer Nature Switzerland AG 2019
K. S. Mohamed, *The Era of Internet of Things*,
https://doi.org/10.1007/978-3-030-18133-8_3

Fig. 1 Communication
media classification
according to connectivity

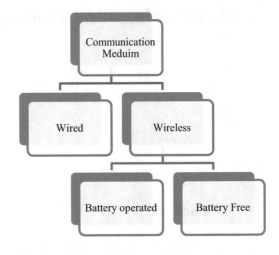

Fig. 2 OSI and TCP/IP
networking models [18]. 7
layers vs. 4 layers.

OSI model	TCP/IP model
7 Application	Application
6 Presentation	
5 Session	
4 Transport	Transport
3 Network	Internet
2 Data link	Network access & physical
1 Physical	

2   IoT Protocol Stack

The Open Systems Interconnection (OSI) model is an ISO-standard abstract model
that describes a stack of seven protocol layers. From the top down, these layers are:
application, presentation, session, transport, network, data link, and physical. TCP/
IP model includes only four layers, merging some of the OSI model layers as shown
in Fig. 2. IoT protocol stack compared to web stack is shown in Fig. 3 [19]. Many
emerging and competing networking technologies are being adopted within the IoT
space [20]. Multiple technologies are offered by different vendors or are aimed at
different vertical markets like home automation, healthcare, or industrial IoT; often
provide alternative implementations of the same standard protocols [18].

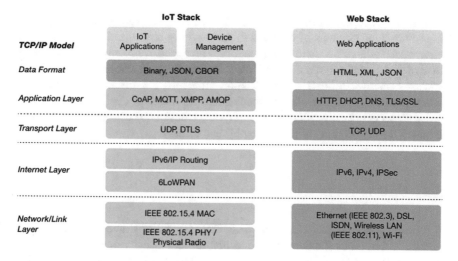

Fig. 3 IoT protocol stack over web stack

3   IoT Network and Link Layer: Wired Communication and Networking

3.1   Ethernet

A system for connecting a number of computer systems to form a local area network, with protocols to control the passing of information and to avoid simultaneous transmission by two or more systems. The first widely used LAN technology was developed in the mid-1970s by researchers at the Xerox Palo Alto Research Center (PARC). It is simpler and cheaper than token LANs and ATM. Every Ethernet network interface card (NIC) is given a unique identifier called a MAC address. The MAC address comprises a 48-bit number. Within the number the first 24 bits identify the manufacturer and it is known as the manufacturer ID or Organizational Unique Identifier (OUI) and this is assigned by the registration authority. Ethernet can't be used for long distance network [21].

Ethernet implements the IEEE 802.3 standard. Not all IoT devices need to be wireless devices that are designed to be stationary. For example, sensor units that are installed within a building automation system can use wired networking technologies like Ethernet.

Ethernet is a wired communication protocol widely used for computer networking. It was first introduced to the market in 1980, and its use got internationally standardized in 1983 as IEEE 802.3. Physically, it uses coaxial cables, twisted pairs, or optical fibers, with networking speeds ranging from 10 Mbits/s up to 100 Gbits/s, and the speed is expected to rise even up to 400 Gbits/s by 2018. Ethernet is usually used in local area network (LAN), in addition to metropolitan area network (MAN) and wide area network (WAN). In OSI model, Ethernet is included in the data link layer.

Due to its support for high-speed communications, Ethernet is ideal for applications with huge amounts of data and those requiring high speed. In addition, it is convenient for high bandwidth applications. Furthermore, Ethernet cables are ideal to transport data to very far destinations. However, Ethernet also suffers from disadvantages relative to other communication protocols. For instance, Ethernet is a wired protocol, which makes it inconvenient for wireless applications. Being a wired protocol not only requires direct physical connection between nodes, but also it makes the connection vulnerable to physical damage.

3.2 USB

It is a representative peripheral interface. USB stands for Universal Serial Bus. It provides a serial bus standard for connecting devices, usually to a computer, but it also is in use on other devices such as set-top boxes, game consoles, and PDAs. Three generations of USB are available: USB 1.0, USB 2.0, and USB 3.0 [22].

4   IoT Network and Link Layer: Wireless Communication and Networking

One of the greatest challenges is choosing the wireless connectivity technology. It is clear that there is no "one size fits all" when it comes to wireless connectivity. Connectivity models are either device to device or device to cloud. The

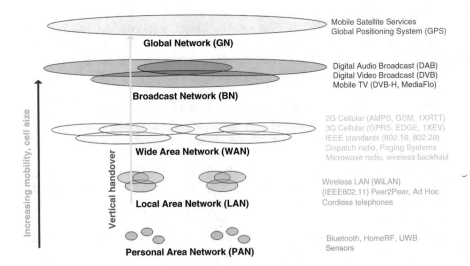

Fig. 4 Communication classifications according to distance

communication layer is considered as the backbone of the IoT systems. It is the main channel between the application layer and different operating activities in the IoT system. The whole physical system is loaded with amounts of data and information that need to be shared with other nodes. Therefore, it is needed to set up a suitable connection network among these nodes through a communication protocol. The communication could be wire-connected or wireless based on the protocol defined by the designer. According to the distance, communication protocols are divided into three main categories as shown in Fig. 4 [17, 23–55]:

1. Short distance communication (PAN): In local areas, numerous data protocols are capable of managing the data flow between local nodes, such as Wi-Fi, which is a wireless networking technology that makes use of radio frequencies to send and receive data, and RFID, which is commonly used in business applications to monitor supply chains especially in manufacturing and retail industries [98].
2. Medium distance communication (LAN): Connecting multiple gateways requires communication protocols for medium distance. Ethernet is a good example because it manages connections between the connected gateways through installed wired systems.
3. Long distance communication (WAN): The existence of active support of satellite networks offers huge communication ranges. Also, supplying the mobile phones with Fifth Generation (5G) communication protocol enhances the communication speed over a very long distance. Since billion of devices around the world communicate together, IoT can easily spread all over the world.
4. Broadcast network (BN)
   A broadcast network is a group of radio stations, television stations, or other electronic media outlets, that form an agreement to air, or broadcast, content from a centralized source.
5. Global network (GN)
A global network is any communication network which spans the entire Earth.

Building a good IoT system requires to connect a very large number of local gateways in a heterogeneous environment through multi-modal technologies and different lightweight protocols such as 5G and to use low-power systems as much as possible. Moreover, rapidly manage the traffic flow of data and enable real-time decision actions and maintain a secure connection for information. In addition, having unlimited addressing capability. The main communication protocols are discussed below.

4.1 Personal Area Network (PAN)

4.1.1 Bluetooth

Bluetooth is a wireless communication protocol widely used nowadays to connect devices together. It was first introduced in 1989 by Nick Rydbeck and Johan Ullman. It uses electromagnetic (EM) waves with frequencies ranging mainly between

2.4 GHz and 2.485 GHz. It is based on a master-slave configuration in which communication is established between a master and up to seven slaves maximum. The latest version is Bluetooth 5, which supports higher Bluetooth speed connections (2 Mbits/s) and further range (more than 300 m). Bluetooth 5 also contains features that support IoT, such as coded communication and forwarded error correction.

There are many advantages for Bluetooth that makes it suitable for IoT use. First of all, it is a wireless protocol, and thus it supports wireless applications such as wearable devices. Secondly, Bluetooth is a low-power protocol, which is also ideal for IoT. On the other hand, Bluetooth limits the number of slave devices connected to a master device (7 devices maximum). In addition, power transmission limits the maximum distance at which a connection could be established, which creates a tradeoff between power consumption and furthest distance for communication [8, 9, 56–61]. Figure 5 shows Bluetooth devices example.

4.1.2   ZigBee

ZigBee is a communication technology for data transfer in wireless networks. It offers low-power connection. In addition, it is designed for multi-channel control systems, alarm systems, and lighting control. Furthermore, ZigBee is more economical than Wi-Fi and Bluetooth as it consumes less power. Moreover, it ensures that networks remain operable in conditions of constantly changing qualities between communication nodes [55]. On the other hand, ZigBee has a low bit transfer rate that only reaches hundreds of kilobits per second; the maximum bit rate transfer for ZigBee is 250 Kbit/s. ZigBee is commonly used for applications that require low data transfer rate and low power consumption. Unfortunately, ZigBee is not widely known like Wi-Fi and Bluetooth as it is often embedded inside systems, and it is not visible. There are three types of ZigBee device: Coordinator, Router, End Device as shown in Figure 6.

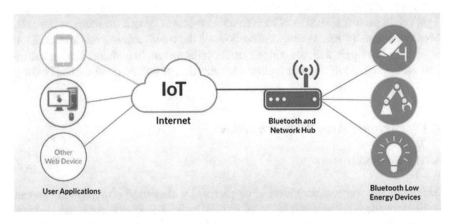

Fig. 5 Bluetooth devices

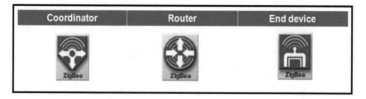

Fig. 6 ZigBee devices

4.1.3   Z-Wave

Z-Wave is a low-power RF communications technology that is primarily designed for home automation for products such as lamp controllers and sensors among many others. Optimized for reliable and low-latency communication of small data packets with data rates up to 100kbit/s, it operates in the sub-1GHz band and is impervious to interference from WiFi and other wireless technologies in the 2.4-GHz range such as Bluetooth or ZigBee. It supports full mesh networks without the need for a coordinator node and is very scalable, enabling control of up to 232 devices. Z-Wave uses a simpler protocol than some others, which can enable faster and simpler development, but the only maker of chips is Sigma Designs compared to multiple sources for other wireless technologies such as ZigBee and others [62].

4.1.4   WSN (Wireless Sensor Network)

A wireless sensor network (WSN) is a collection of distributed sensors that monitor physical or environmental conditions, such as temperature, sound, and pressure. Data from each sensor passes through the network node-to-node. The networks typically run low-power devices. It may consist of one or more sensors and could be different type of sensors.

A sensor node is typically a tiny electronic device equipped with: Low-cost microcontroller with very low power consumption, one or more sensors that gather data of interest from the surroundings, limited battery which is assumed to neither be replaced nor recharged as the power source, flash memory, and transceiver that uses Radio Frequency (RF) for communication with other nodes. Laser and infrared also can be used for communication instead of RF in some WSN applications, but they are sensitive to atmospheric conditions, and also require line of sight.

Fig. 7 shows WSN example. Waspmote [63] and Meshlium [64] are examples for WSN products. The central components of the sensor nodes are [65]:

- Sensor: It encloses an embedded chip for sensing vital medical signs from the body of patient.
- Microcontroller: It controls the function of the other components and accomplishes local data processing including data compression.
- Memory: It temporally stores the sensed data obtained from the sensor nodes.

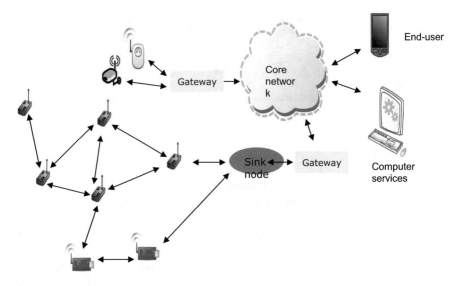

Fig. 7 WSN example

- Radio transceiver: It communicates the nodes and allows physiological data to be wirelessly sent/received.
- Power supply: It is used to supply the sensor nodes by the required powered through batteries.
- Signal conditioning: It amplifies and filters the physiological sensed data to suitable levels for digitization.
- Analog to digital converter (ADC): It produces digital signals from the analog ones to allow further required processes.

A sensor node mainly uses energy in two conditions. The first condition is when a sensor node performs tasks such as transmitting data packets, receiving data packets, processing requests, which are called useful tasks. While operations such as transmission and reception of control packets with neighboring nodes for upper or lower layer head selection or for choosing better forwarder, retransmissions, overhearing, processing redundant packets, and similar other operations come under second condition, which is considered a waste of energy.

4.1.5 Radio Frequency Identification Tags (RFID)

Radio frequency identification sensors play an important role in transmitting and receiving data. A RFID tag carries data and sends them in radio waves to a RFID reader that reads them. It does not require close communication between the tag and the reader; instead it can identify itself from a distance without any human factor. There are two common RFID configurations: near and far. The former configuration uses a RFID reader that has a coil through which alternating current passes and hence generates magnetic field. The tag must have a smaller coil that generates potential due to the changes in the magnetic field. It is then coupled with a capacitor to power up a tag chip. Meanwhile, the latter configuration has a reader and a tag with dipole antenna in which EM waves propagate.

4.1.6   NFC (Near Field Communication)

Near Field Communication (NFC) is a set of short-range wireless technology at 13.56 MHz, typically requiring a distance of 4 cm. NFC technology makes life easier and more convenient for consumers around the world by making it simpler to make transactions, exchange digital content, and connect. NFC is a very short range wireless communication technology that enables the data transmission among devices by touching them together or bringing them together no more than a few inches. NFC uses similar technology principles in RFID. However, it is not only used for identification but also for more elaborate two-way communication. NFC has a tag that can contain small amount of data.

This tag can be read only (similar to RFID tags for identification purposes) or can be rewritable and be altered later by the device. There are three main operating modes for NFC: card emulation mode (passive mode), reader/writer mode (active mode), and peer-to-peer mode. NFC technology is extensively used in mobile phones, industrial applications, and contactless payment systems. Similarly, NFC makes it easier to connect, commission, and control IoT devices in different environments like home, factory, and the work. NFC supports P2P network topology.

4.2   Local Area Network (LAN)

4.2.1   Wi-Fi

Wi-Fi is an easy and inexpensive communication protocol that connects electronic devices to the Internet using a wireless router. This router receives the signal and sends the information to the Internet using Ethernet. Wi-Fi is used mainly as a replacement for the high-speed cables in local areas. The latest version of Wi-Fi nowadays is 802.11 ac, which has super high speed ranging from 433 Mbps up to gigabytes per second, and offering wide ranges of bandwidth (80 MHz and 160 MHz). The range of Wi-Fi reaches 100 meters. There are many IEEE standards for WiFi. 802.11ax also known as "high-efficiency wireless" will be commonly referred to as **Wi-Fi 6.** Technically, Wi-Fi 6 will have a single-user data rate that is 37% faster than 802.11 ac, but what's more significant is that the updated specification will offer four times the throughput per user in crowded environments, as well as better power efficiency which should translate to a boost in device battery life [66].

4.2.2 WiMAX

WiMAX is not a replacement technology to WiFi—instead, while WiFi is the de-facto global standard for wireless interconnection of end-user devices, WiMAX has addressed a specific technical deficiency of WiFi for interconnection of multiple sites. The main drawback of WiFi technology for a point-to-multipoint connection is that it is a connectionless type of protocol named CSMA/CA (carrier sense multiple access with collision avoidance), which means that as in WiFi networks all the devices of the network share the same frequency channel, to prevent collision in data transmissions, each device "listens" to make sure no other device is transmitting and then it transmits its data. That is, there is no centralized management in the network. While this makes the network setup very simple and straightforward (which is a benefit for end-user devices), it creates major problems in larger networks especially when the distances are increased.

WiMAX has mainly addressed this issue by adopting a fully connection-based protocol, which uses a scheduling algorithm. Unlike a WiFi network, in WiMAX you should define and set up each subscriber station (SS) on the base station including specifying what bandwidth each SS should be given. By doing this, the base station knows the exact number of subscriber stations and allocates a time slot (access slot) to each. This protocol synchronizes the transmission of data between all the stations on the network and totally eliminates the collision issues of a WiFi network. This enables efficient and reliable connection of as many as 80 subscribers on a WiMAX network with guaranteed QoS (Quality of Service), while on an outdoor WiFi network, adding more than 10 CPEs would cause great deficiency with unpredictable quality of service [67].

4.3 Wide Area Network (WAN)

4.3.1 Fifth Generation (5G)

5G is the most recent data protocol used in smartphones and expected to be more than a new generation. It will introduce a new era of connectivity as it will offer speeds of more than 100 megabits per second, one more data bandwidth, and less delay due to built-in computing intelligence that handles data very efficiently. So, it will connect billions of devices in the fastest, most reliable and most efficient ways. As a result, 5G is expected to take IoT to a new revolutionary level and extend its use more and more [68].

5G doesn't just deliver peak speeds in ideal conditions. The technology offers super high speeds that are reliable and consistent, even indoors or in congested areas. 5G can support a massive increase in connected devices. Ericsson forecasts 1 billion 5G subscriptions by 2023. Phones today have an annoying lag between when you send a request for a website or video and when the network responds. With 5G, that'll be reduced to 1 millisecond. That's 400 times faster than the blink of an eye. It's so fast, some companies see it opening up the possibility of remote surgery [69].

When we look at the needs and requirements of many of the IoT use cases, we can see that cellular has a fairly unique combination of characteristics that makes it a better choice for IoT. When comparing inherent capabilities of cellular IoT with other LPWA technologies like Sigfox and Lora, cellular IoT offers better performance in terms of unmatched global coverage, quality of service, scalability, and flexibility of handling a comprehensive range of use cases. Cellular IoT has also quite extensive ecosystem based on standardization, supported by many global top operators, device maker, chipset and module vendor and network vendors. Most importantly, unlicensed solutions can't guarantee reliability and security as they have to compete for bandwidth with other wireless networks. Last but not least, cellular IoT has the advantage of total cost of ownership thanks to an easy software upgrade on existing networks, and full multiplex and reuse of existing network resource, which will help to reduce operational costs in area like service provisioning, monitoring, and billing [70]. Comparison between 1G, 2G, 3G, 4G, and 5G is shown in Table 1 [71–73].

4.3.2   LTE-E and NB-IoT

The Internet of Things (IoT) will massively expand the use of cellular communications beyond smartphones and tablets to an extraordinary range of applications and connected devices—including many products that haven't yet been invented. Although many IoT applications only need short-range wireless connectivity, a significant number will require longer-range connections. Examples include utility meters, sensors that monitor farm fields, or telematics modules tracking trucks on cross-country routes. That's where cellular IoT comes in, and specifically the two cellular IoT standards expected to dominate the market: NB-IoT and LTE-M [74, 75].

LTE-M (Long Term Evolution for Machines) and NB-IoT (narrowband Internet of Things) are standards created by 3GPP, the standards organization responsible

Table 1 Comparison between 1G, 2G, 3G, 4G, and 5G

	1G	2G	3G	4G	5G
Period	1980–1990	1990–2000	2000–2010	2010–2020	2020–2030
Bandwidth	150/900 MHz	900 MHz	100 MHz	100 MHz	1000× BW per unit area
Frequency	30 KHz analog signal	1.8 GHz (digital)	1.6–2.0 GHz	2–8 GHz	3–300 GHz
Data rate	2 kbps	64 kbps	144 kbps–2Mbps	100Mbps–1Gbps	>1Gbps
Characteristic	First wireless communication	Digital	Digital broadband	High speed	–
Technology	Analog cellular	Digital cellular (GSM)	CDMA, UMTS, EDGE	LTE, WiFi	–

for LTE and 5G. Both LTE-M and NB-IoT fall under the category of machine-to-machine (M2M) communication. They help enable applications such as smart cities, environmental monitoring, asset tracking, and more. LTE-M and NB-IoT are different because they're specifically designed and optimized for IoT devices that communicate small amounts of data over long periods of time. So they're simpler than other cellular standards, with much less overhead. NB-IoT and LTE-M are the natural successors to older cellular standards for existing applications, and they'll also drive development of completely new applications [76–80].

4.3.3 Sigfox: WiMAX for IoT

An alternative wide-range technology is Sigfox, which in terms of range comes between WiFi and cellular. It uses the ISM bands, which are free to use without the need to acquire licenses, to transmit data over a very narrow spectrum to and from connected objects. The idea for Sigfox is that for many M2M applications that run on a small battery and only require low levels of data transfer, then WiFi's range is too short while cellular is too expensive and also consumes too much power. Sigfox uses a technology called Ultra Narrow Band (UNB) and is only designed to handle low data-transfer speeds of 10 to 1000 bits per second. It consumes only 50 microwatts compared to 5000 microwatts for cellular communication, or can deliver a typical stand-by time 20 years with a 2.5 Ah battery while it is only 0.2 years for cellular [81].

Already deployed in tens of thousands of connected objects, the network is currently being rolled out in major cities across Europe, including ten cities in the UK for example. The network offers a robust, power-efficient and scalable network that can communicate with millions of battery-operated devices across areas of several square kilometers, making it suitable for various M2M applications that are expected to include smart meters, patient monitors, security devices, street lighting, and environmental sensors. The Sigfox system uses silicon such as the EZRadioPro wireless transceivers from Silicon Labs, which deliver industry-leading wireless performance, extended range, and ultra-low power consumption for wireless networking applications operating in the sub-1GHz band.

4.3.4 Neul

Similar in concept to Sigfox and operating in the sub-1GHz band, Neul leverages very small slices of the TV white space spectrum to deliver high-scalability, high-coverage, low-power and low-cost wireless networks. Systems are based on the Iceni chip, which communicates using the white space radio to access the high-quality UHF spectrum, now available due to the analog to digital TV transition. The communications technology is called Weightless, which is a new wide-area wireless networking technology designed for the IoT that largely competes against existing GPRS, 3G, CDMA, and LTE WAN solutions. Data rates can be anything from a few bits per second up to 100 kbps over the same single link; devices can consume as little as 20–30 mA from 2×AA batteries, meaning 10–15 years in the field [82].

4.3.5   LoRaWAN

Similar in some respects to Sigfox and Neul, LoRaWAN targets wide area network (WAN) applications and is designed to provide low-power WANs with features specifically needed to support low-cost mobile secure bi-directional communication in IoT, M2M and smart city and industrial applications. Optimized for low power consumption and supporting large networks with millions and millions of devices, data rates range from 0.3 to 50 kbps. Figure 8 shows the communication protocol and system architecture of LoRaWAN [83–90].

LoRa is developed by Semtech Inc. as a proprietary physical layer protocol to provide low-power and long-distance communication up to 20 km (unobstructed line-of-sight) by using a special radio modulation technique Chirp Spread Spectrum (CSS) as depicted in Fig. 9.

4.4   Broadcast Network (BN)

4.4.1   Digital Video Broadcast (DVB)

Digital Video Broadcasting (DVB) is being adopted as the standard for digital television in many countries. The DVB project was an industry led consortium of over 270 television broadcasting associated companies worldwide. The DVB standard offers many advantages over the previous analog standards and has enabled television to make a major step forwards in terms of its technology. DVB is now one of the success stories of modern broadcasting. The take-up has been enormous and it is currently deployed in over 80 countries worldwide, including most of Europe and also within the USA. It offers advantages in terms of far greater efficiency in terms of spectrum usage and power utilization as well as being able to affect considerably more facilities, the prospect of more channels, and the ability to work alongside existing analog services [92].

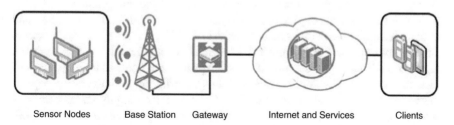

| Sensor Nodes | Base Station | Gateway | Internet and Services | Clients |

Fig. 8 The communication protocol and system architecture of LoRaWAN

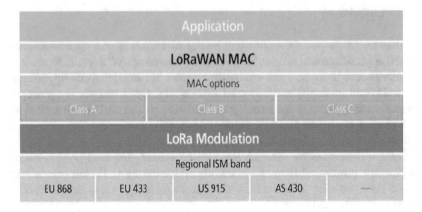

Fig. 9 A layered description of LoRa technology [91]

4.5 Global Network (GN)

4.5.1 Satellites Networks

Satellites have a great role behind the growth of IoT. They enable the development of consumer-centric services that gives good experience to the user. They supply IoT with huge advantages that cannot be offered by Wi-Fi deployments, Bluetooth, or terrestrial GSM (Global System for Mobile Communications) networks like coverage. Satellite networks serve to cover a huge range, therefore IoT is expected to encompass billions of devices around the world even in remote locations. As a result, using active support of satellite networks, such as the L-band services provided by Thuraya, will ensure the ultimate success of IoT expansion [93].

5 IoT Internet Layer

5.1 IPV6/6LowWPAN

Thanks to its large address space, IPv6 enables the extension of the Internet to any device and service. Experiments have demonstrated the successful use of IPv6 addresses to large scale deployments of sensors in smart buildings, smart cities, and even with cattle. IPv6 delivers other benefits in addition to a larger addressing space, for example, permitting hierarchical address allocation techniques that limit the expansion of routing tables, simplified and expanded multicast addressing, and service delivery optimization. Device mobility, security, and configuration aspects have been considered in the design of IPv6. IPv6 offers a highly scalable address scheme. 6LowWPAN is IPV6 over low-power wireless personal area network [42]. IPv6 is considered the best protocol for communication in the IoT domain because of its scalability and stability.

6   IoT Application Layer

6.1   *CoAP*

CoAP is designed to suit the energy constraints and low processing power of IoT devices. CoAP describes a web transfer protocol bound to the user datagram protocol most suited to IoT applications. CoAP changes the functionalities of the hyper text transfer protocol, based on the requirements of IoT devices. CoAP is divided into a messaging sublayer and a request/response sublayer. The messaging sublayer identifies duplications and provides reliable communication. The request/response sublayer offers services to users. CoAP uses four types of messages: confirmable, non-confirmable, reset, and acknowledgement. There are four types of response modes in CoAP. The first is the separate response mode that is used when the server is required to wait for a period of time before replying to the client. Under the non-confirmable response mode, the client sends the data without waiting for an acknowledgement. Just like HTTP, CoAP uses methods like GET, PUT, POST, and DELETE. CoAP includes features such as resource observation, block-wise resource transport, and resource discovery [15]. CoAP layer as a part of the seven layers model is shown in Fig. 10.

6.2   *MQTT*

The message queue telemetry transport (MQTT) is a messaging protocol that enables connections between IoT devices. It uses a routing mechanism that makes it an optimal connection protocol for M2M. MQTT has three components: a subscriber, a broker, and a publisher [15]. The IoT device willing to communicate can

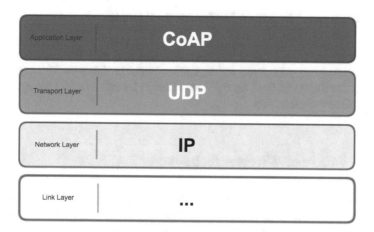

Fig. 10 CoAP layer as a part of the seven layers model

Fig. 11 MQTT protocol
stack

Table 2 Comparison of MQTT and CoAP.

	TCP/UDP	Architecture	Header size (Bytes)	Length (Bytes)
MQTT	TCP	Pub/Sub	2	5
CoAP	UDP	Req/Res	4	20

register as a subscriber for certain topics and is informed by the broker when publishers publish topics of interest. The publisher is responsible for transmitting this information to interested IoT devices through the broker, which is responsible for security and providing authentication [16]. Protocol stack is shown.

Figure 11. The publisher subscriber model allows MQTT clients to communicate one-to-one, one-to-many, and many-to-one. A comparison of MQTT and CoAP is shown in Table 2.

7 Comparison between Different IoT Protocols

Table 3 provides a comparison between different IoT protocols in terms of frequency, range, encryption, modulation, applications, and data rate. This comparison aims at presenting guidelines for the researchers to be able to select the right protocol for different applications.

8 The Future of Wireless Technology

Future wireless networks will support 100 Gbps communication between people, devices, and IoT with high reliability and uniform coverage indoors and outdoors. The shortage of spectrum to support such systems will be alleviated by advances in

Table 3 Comparison between different IoT protocols

	Frequency	Range	Data rate	Power	Encryption	Modulation	Preamble [94]	Application
Bluetooth	2.4GHz	50–150 m	1Mbps	30 mA	AES	GPSK	4 bytes "01010101"	Wireless headsets, audio
ZigBee	2.4GHz	10–100 m	250 kbps	30 mA	AES	BPSK	4 bytes "01010101"	Home automation
Z-Wave	900 MHz	30 m	100 kbps	2.5 mA	AES	BFSK	m bytes "01010101"	Home automation
WIFI	2.4GHz and 5GHz	50 m	600 mbps	High	AES	–	Configuration specific	M2M
Cellular (4G)	2.1 GHz	200 km	3–10 mbps	High	RC4	GFSK	–	M2M
RFID	Many	200 m	4 mbps	Low power	RC4	FSK, PSK	–	Tracking, inventory, access
NFC	13.56 MHz	10 cm	100–420 kbps	50 mA	RSA	ASK	–	Payment, access
Sigfox	900 MHz	30-50 km	1 kbps	100 mW	AES	GFSK	4 bytes unknown	Street lighting
Neul	900 MHz	10 km	100 kbps	100 mW	AES	GFSK	–	Street lighting
LoRaWAN	Many	2–5 km	50 kbps	100 mW	AES	GFSK	Sequence of 1 s	Low-power connectivity
6LoWPAN	2.4GHz	100 m	250 kbps	1–2 years lifetime on batteries	AES	BPSK		Home automation

massive **MIMO** and **mmW** technology as well as **cognitive radios**. Breakthrough energy-efficiency architectures, algorithms, and hardware will allow wireless networks to be powered by tiny batteries, energy harvesting, or **over-the-air power transfer**. Finally, **new communication systems** based on biology and chemistry to encode bits will enable a wide range of new micro- and macroscale applications. However, there are many technical challenges that must be overcome in order to make this vision a reality.

9   Conclusions

This chapter presents a survey of the latest requirements of the IoT with associated communication technologies. Moreover, it provides a comparison between different IoT protocols in terms of frequency, range, encryption, modulation, applications, QoS, and data rate. This comparison aims at presenting guidelines for the researchers to be able to select the right protocol for different applications.

References

1. Network Connectivity for IoT. Retrieved from https://6s062.github.io/6MOB/2017/materials/lec5-IOTx-WirelessNetworkConnectivity.pdf.
2. Santitoro, R. *Metro Ethernet services—A technical overview*. Metro Ethernet Forum.
3. Tonner, D. (2007) The bluetooth blues | information age. Web.archive.org. [Online]. Retrieved July 19, 2017, from https://web.archive.org/web/20071222231740/, http://www.informationage.com/article/2001/may/the_bluetooth_blues
4. BR/EDR: Point-to-Point | Bluetooth Technology Website. Bluetooth.com, 2017. [Online]. Retrieved July 19, 2017, from https://www.bluetooth.com/what-is-bluetooth-technology/how-it-works/br-edr.
5. Bluetooth basics—learn.sparkfun.com. Learn.sparkfun.com, 2017. [Online]. Retrieved July 18, 2017, from https://learn.sparkfun.com/tutorials/bluetooth-basics/how-bluetooth-works.
6. Nield, D. (2017). *Bluetooth 5: Everything you need to know*. TechRadar. [online]. Retrieved July 18, 2017, from http://www.techradar.com/news/networking/bluetooth-5-everything-you-need-to-know-1323060.
7. Sims, G. (2017). *The truth about Bluetooth 5 - Gary explains*. Android Authority. [Online]. Retrieved August 8, 2017, from http://www.androidauthority.com/bluetooth-5-speed-range-762369/.
8. Mohammed, K. S. (2009). FPGA implementation of PPM I-UWB baseband transceiver. In *Proceedings of the European computing conference*. Boston, MA: Springer.
9. Salah, K. (2008). Design and FPGA implementation of non-data aided timing and carrier recovery techniques for EDR Bluetooth standard. *Signal processing algorithms, architectures, arrangements, and applications (SPA), 2008*. IEEE.
10. Salah, K. (2006). FPGA implementation of Bluetooth 2.0 transceiver. *Proceedings of the 5th WSEAS international conference on system science and simulation in engineering*. World Scientific and Engineering Academy and Society (WSEAS).
11. What is WiFi and How Does it Work? CCM, 2017. [Online]. Retrieved July 18, 2017, from http://ccm.net/faq/298-what-is-wifi-and-how-does-it-work.

12. Lendino, J. (2016). What is 802.11ac Wi-fi, and how much faster than 802.11n is it? - ExtremeTech", ExtremeTech. [Online]. Retrieved July 24, 2017, from https://www.extreme-tech.com/computing/160837-what-is-802-11ac-and-how-much-faster-than-802-11n-is-it.

13. *Explaining wireless sensor nodes: Zigbee vs. WiFI.* YouTube, 2017. [Online]. Retrieved July 18, 2017, from https://www.youtube.com/watch?v=buV11ZPJ7MQ.

14. *CCTV Institute | CCTV Surveillance Smart-homes Home Automation Zigbee.* CCTV Institute, 2017. [Online]. Retrieved July 18, 2017, from http://cctvinstitute.co.uk/zigbee/.

15. Shelby, Z., Hartke, K., Bormann, C. and Frank, B. (2013). *Constrained Application Protocol (CoAP), draft-ietf-corecoap-18, Internet Eng.* Fremont, CA: Task Force (IETF).

16. Locke, D. (2010). *MQ Telemetry Transport (MQTT) v3. 1 Protocol Specification.* Markham, ON: IBM Developer Works, Tech. Lib.

17. Tan, L. & Wang, N. (2010). *Future internet: The internet of things.* Advanced computer theory and engineering (ICACTE), 2010 3rd International Conference on: V5–376.

18. Retrieved from https://developer.ibm.com/articles/iot-lp101-connectivity-network-protocols/

19. Gorrepotu, R. (2018). Sub-1GHz miniature wireless sensor node for IoT applications. *Internet of Things, 1–2,* 27–39. Elsevier.

20. Pokhrel, S. R., & Williamson, C. (2018). Modeling compound TCP over WiFi for IoT. *IEEE/ACM Trans. Netw., 26,* 864–878.

21. Retrieved from *www.fujitsu.com/downloads/TEL/fnc/pdfservices/ethernet-prerequisite.pdf*

22. Retrieved from https://www.computer-solutions.co.uk/info/Embedded_tutorials/usb_tutorial.htm

23. Strategy, I. & Unit, P. (2005). *ITU Internet Reports 2005: The internet of things.* Geneva: International Telecommunication Union (ITU).

24. Li, X., Xuan, Z., & Wen, L. (2011). Research on the architecture of trusted security system based on the internet of things. *Intelligent Computation Technology and Automation (ICICTA), 2011 International Conference on.* 1172–1175.

25. Porkodi, R., & Bhuvaneswari, V. (2014). The internet of things (IoT) applications and communication enabling technology standards: An overview. *Intelligent Computing Applications (ICICA), 2014 International Conference on.* 324–329.

26. Samie, F., Bauer, L., & Henkel, J. (2016). IoT technologies for embedded computing: A survey. *Hardware/software Codesign and system synthesis (CODES+ ISSS), 2016 international conference on.* 1–10.

27. Salman, T. (2015). Internet of things protocols and standards. Affairs, M. Of E. N.D. 2015. Internet of things in the Netherlands applications trends and potential impact on radio spectrum.Startix.

28. Paavola, M. (2007). *Wireless technologies in process automation-review and an application example.* Control Engineering Laboratory, University of Oulu.

29. Le, A., Loo, J., Lasebae, A., Aiash, M., & Luo, Y. (2012). 6LoWPAN: A study on QoS security threats and countermeasures using intrusion detection system approach. *International Journal of Communication Systems, 25*(9), 1189–1212.

30. Martha Zemede, K. T. (2015). *Explosion of the internet of things: What does it mean for wireless devices?.* Keysight Technologies.

31. Goursaud, C., & Gorce, J.-M. (2015). *Dedicated networks for IoT: PHY/MAC state of the art and challenges.* EAI endorsed transactions on internet of things.

32. Gomez, C., & Paradells, J. (2010). Wireless home automation networks: A survey of architectures and technologies. *IEEE Communications Magazine, 48*(6), 92.

33. Rathnayaka, A. D., Potdar, V. M., & Kuruppu, S. J. (2011). Evaluation of wireless home automation technologies. *Digital Ecosystems and Technologies Conference (DEST), 2011 Proceedings of the 5th IEEE International Conference on:* 76–81.

34. Aragues, A., Martinez, I., Del Valle, P., Muñoz, P., Escayola, J., & Trigo, J. D. (2012). Trends in entertainment, home automation and e-health: Toward cross-domain integration. *IEEE Communications Magazine, 50*(6), 160.

35. López, P., Fernández, D., Jara, A. J. & Skarmeta, A. F. (2013). Survey of internet of things technologies for clinical environments. *Advanced Information Networking and Applications Workshops (WAINA), 2013 27t International Conference on*: 1349–1354.
36. Tabish, R., Mnaouer, A. B., Touati, F. & Ghaleb, A. M. (2013). A comparative analysis of BLE and 6LoWPAN for U-HealthCare applications. *GCC Conference and Exhibition (GCC), 2013 7th IEEE*. 286–291.
37. Al-Fuqaha, A., Guizani, M., Mohammadi, M., Aledhari, M., & Ayyash, M. (2015). Internet of things: A survey on enabling technologies, protocols, and applications. *IEEE Communications Surveys \& Tutorials, 17*(4), 2347–2376.
38. Kuzlu, M., Pipattanasomporn, M. & Rahman, S. (2015). Review of communication technologies for smart homes/building applications. *Innovative Smart Grid Technologies-Asia (ISGT ASIA), 2015 IEEE*: 1–6.
39. Samuel, S. S. I. (2016). A review of connectivity challenges in IoT-smart home. *Big data and Smart City (ICBDSC), 2016 3rd MEC international conference on*: 1–4.
40. Raza, U., Kulkarni, P., & Sooriyabandara, M. (2017). *Low power wide area networks: An overview*. IEEE Communications Surveys & Tutorials.
41. Frantz, T. L. & Carley, K. M. (2005). *A formal characterization of cellular networks*.
42. Hossen, M., Kabir, A., Khan, R. H., Azfar, A. & others. 2010. *Interconnection between 802.15. 4 devices and IPv6: implications andexisting approaches*. arXiv preprint arXiv:1002.1146.
43. Azamuddin Bin Ab Rahman, R. J. (2015). Comparison of Internet of Things (IoT) Data Link Protocols.
44. Alliance, L. 2015. A technical overview of LoRa and LoRaWAN. White Paper, November.
45. Shreya Shah, T. M. n.d. Security of NFC Data. *International Journal of Advanced Research in Computer Science and Software Engineering, 6*, (ISSN: 2277 128X).
46. Hughes, J., Yan, J., & Soga, K. (2015). Development of wireless sensor network using bluetooth low energy (BLE) for construction noise monitoring. *International Journal on Smart Sensing and Intelligent Systems, 8*(2), 1379–1405.
47. Ahmad, A. (2005). *Wireless and mobile data networks*. Wiley.
48. Gomez, C., Oller, J., & Paradells, J. (2012). Overview and evaluation of bluetooth low energy: An emerging low-power wireless technology. *Sensors, 12*(9), 11734–11753.
49. Sanchez-Iborra, R., & Cano, M.-D. (2016). State of the art in LP-wan solutions for industrial IoT services. *Sensors, 16*(5), 708.
50. Cerruela Garcia, G., Luque Ruiz, I., & Gómez-Nieto, M. Á. (2016). State of the art, trends and future of Bluetooth low energy, near field communication and visible light communication in the development of smart cities. *Sensors, 16*(11), 1968.
51. Frenzel, L. (2012). *The fundamentals of short-range wireless technology*. Electronic Design.
52. Alarcon-Aquino, V., Dominguez-Jimenez, M., & Ohms, C. (2008). Desing and implementation of a security layer for RFID systems. *Journal of Applied Research and Technology, 6*(2), 69–82.
53. Amin, M., Reaz, M., Jalil, J., & Rahman, L. (2012). Digital modulator and demodulator IC for RFID tag employing DSSS and barker code. *Journal of Applied Research and Technology, 10*(6), 819–825.
54. Friess, P. (2013). *Internet of things: Converging technologies for smart environments and integrated ecosystems*. River Publishers.
55. Lu, C.-W., Li, S.-C. & Wu, Q. 2011. Interconnecting ZigBee an 6LoWPAN wireless sensor networks for smart grid applications. *Sensing Technology (ICST), 2011 Fifth International Conference on*: 267–272.
56. Salah, K. (2006). FPGA implementation of Bluetooth 2.0 transceiver. *Proceedings of the 5th WSEAS international conference on system science and simulation in engineering*. World Scientific and Engineering Academy and Society (WSEAS), 2006.
57. Chang, K. H. (2014). Bluetooth: A viable solution for IoT? [industry perspectives]. *IEEE Wireless Communications, 21*(6), 6–7.

58. Pandya, H. B., Champaneria, T. A. Internet of things: Survey and case studies. *2015 international conference on electrical, electronics, signals, communication and optimization (EESCO), Jan 2015*, pp. 1–6.

59. ABI Research. *Bluetooth 5 evolution will lead to widespread deployments on the IoT landscape*. London, July 2016.

60. Rappaport, T. S. (2002). *Wireless communications: Principles and practice*. Prentice Hall.

61. Bluetooth Special Interest Group. (2016). Bluetooth Core Specifications. Retrieved from https://www.bluetooth.com/specifications/bluetooth-core specification.

62. Retrieved from https://z-wavealliance.org/

63. Retrieved from http://www.libelium.com/products/waspmote/

64. Retrieved from http://www.libelium.com/products/meshlium/

65. Safeer, K. P., Gupta, P., Shakunthala, D. T., Sundersheshu, B. S., & Padaki, V. C. (2008). Wireless sensor network for wearable physiological monitoring. *Journal of Networks, 3*(5), 21–29.

66. Retrieved from https://www.techspot.com/article/1769-wi-fi-6-explained/?fbclid=IwAR3vs-LO-p6CT0WXD5fB7QOoNWOTSLCI10L8LdnrL5Tkt0l8ldtpuySc4-I

67. Retrieved from http://www.vizocom.com/blog/wimax-differ-wifi/

68. Wang et al. (2014). Cellular architecture and key technologies for 5G wireless communication networks. *IEEE Communications Magazine, 52*(2), 122–130.

69. Akpakwu, et al. (2018). A survey on 5G networks for the internet of.Ings: Communication technologies and challenges. *IEEE Access, 6*, 3619–3647.

70. Palattella, M. R., Dohler, M., Grieco, A., Rizzo, G., Torsner, J., Engel, T., & Ladid, L. (Mar. 2016). Internet of things in the 5G era: Enablers, architecture, and business models. *IEEE Journal on Selected Areas in Communications, 34*(3), 510–527.

71. Retrieved from http://www.futuretimeline.net/blog/2015/01/22.htm#.V9e4TvmLRhE

72. Retrieved from http://www.phonearena.com/news/1G-2G-3G-4G-The-evolution-of-wireless generations_id46952.

73. Retrieved from http://gizmodo.com/what-is-5g-and-how-will-it-make-my-life-better-1760847799

74. Ratasuk, R.; Mangalvedhe, N.; Zhang, Y.; Robert, M.; Koskinen, J.P. Overview of narrowband IoT in LTE Rel-13. *Proceedings of the IEEE conference on standards for communications and networking (CSCN)*, Berlin, Germany, 31 October–2 November 2016; pp. 1–7.

75. Zayas, A.D., & Merino, P. The 3GPP NB-IoT system architecture for the internet of things. *Proceedings of the IEEE International Conference on Communications Workshops (ICC Workshops)*, Paris, France, 21–25 May 2017; pp. 277–282.

76. Chen, M., Miao, Y., Hao, Y., & Hwang, K. (2017). Narrow band internet of things. *IEEE Access, 5*, 20557–20577.

77. Adhikary, A., Lin, X., & Wang, Y. P. E. (2017). Performance evaluation of NB-IoT coverage. *IEEE Symposium on Communications and Vehicular Technology*.

78. Boisguene, R., Tseng, S. C., Huang, C. W., Lin, P. (2017). A survey on NB-IoT downlink scheduling: Issues and potential solutions. *13th Int. Wirel. Commun. Mob. Comput. Conf.*

79. IWCMC. (2017). pp. 547–551, 2017.

80. *NB-IoT vs LoRa technology - which could take gold?* 2016. Retrieved from https://www.lora-alliance.org/lorawan-whitepapers.

81. "Sigfox-Iot-Technology-Overview @ Www.Sigfox.Com." [Online]. Retrieved from https://www.sigfox.com/en/sigfox-iot-technology- overview.

82. Retrieved from https://neul.com/

83. Www.Lora-Alliance.Org. [Online]. Retrieved from https://www.lora-alliance.org/.

84. Se mtech Corporation. LoRa modulation basics, 2015. Retrieved from https://www.semtech.com/technology

85. Bor, M., Roedig, U. (2017). Lo Ra transmission parameter selection. *2017 13th International Conference on Distributed Computing Systems*. pp. 27–34.

86. Robert, J., Heuberger, A. (2017). LPWAN downlink using broadcast transmitters. *IEEE International Symposium on Broadband Multimedia Systems and Broadcasting, BMSB*, 2017.
87. Lo RaWAN TM 101, A technical introduction, 2017. Retrieved from https://www.lora -alliance.org/lorawan-whitepapers "products @ www.semtech.com" [Online]. Retrieved from https://www.semtech.com/products.
88. Marais, J. M., Malekian, R., Abu-Mahfouz, A. M. Lo Ra and LoRaWAN testbeds: A review. *2017 IEEE AFRICON Science Technology Innovation. Africa, AFRICON 2017*, pp. 1496–1501, 2017.
89. J. de Carvalho Silva, J. J. P. C. Rodrigues, A. M. Alberti, P. Solic, and A. L. L. Aquino, LoRaWAN—a low power WAN protocol for internet of things: A review and opportunities. *2017 2nd International Multidisciplinary Conference on Computer and Energy Science*, pp. 1–6, 2017.
90. C. P. San, J. Bergs, C. Hawinkel, and J. Famaey, "Comparison of Lo RaWAN classes and their power consumption," *IEEE Symposium on Communications and Vehicular Technology*, pp. 8–13, 2017.
91. LoRaWAN 1.1 Specification. (2017). Retrieved October 22, 2017, from http://lora-alliance.org/lorawan-for-developers.
92. Retrieved from https://www.electronics-notes.com/articles/audio-video/broadcast-tv-television/what-is-dvb-digital-video-broadcasting-tutorial.php.
93. Retrieved from https://www.orbcomm.com/en/networks/satellite.
94. Narayanan, R. (2018). *Revisiting software defined radios in the IoT era*. ACM.
95. Mohammed, K. S. FPGA implementation of PPM I-UWB baseband transceiver. *Proceedings of the European computing conference*. Boston, MA: Springer, 2009.

IoT Cloud Computing, Storage, and Data Analytics

1 Introduction

In this chapter, we will discuss the basic motivation behind cloud computing and its importance for IoT. Thanks to TCP/IP and HTTP, any client or IoT hardware can talk to any IoT service, no matter which hardware you choose. We will learn how to use the cloud as an intelligent system.

The main value of IoT lies in the data analysis and processing. The larger the data is, the larger the computers storing these data must be. In early 1990s, direct server storage quickly became unmanageable as storage demand increased, and there was no way to pull capacity across multiple servers. Hence, data centers emerged when UNIX servers replaced mainframes for running most business applications in their high-end server environments. A data center is a technical facility that houses an organization's IT operations and equipment.

The efficiency, optimization, reliability, and security are the most critical factors in any data center, because it has the network of all systems and data. Data centers may vary according to their primary function. They have technical spaces and subsystems such as physical security systems, network and IT systems, power resources, environmental control, and performance and operational management [1, 2].

A software defined data center (SDDC) is a data storage facility in which infrastructure elements of the data center are virtualized and delivered as a service. Virtualization is the main algorithm of the SDDC. There are three types of virtualization: network virtualization, storage virtualization, and server virtualization. Network virtualization connects network resources by splitting the bandwidth into several independent channels that can be assigned to a particular server or device in the same time. Storage virtualization turns physical storage from multiple network storage devices into a single storage device managed by a central console. Server virtualization is about providing server resources that contain the identity of individual physical servers, processors, and operating systems, from server users. The new approach is to spare users from managing complicated server-resource details.

© Springer Nature Switzerland AG 2019
K. S. Mohamed, *The Era of Internet of Things*,
https://doi.org/10.1007/978-3-030-18133-8_4

2   Cloud Computing

2.1   Cloud Computing: What?

Cloud computing means storing and accessing data and programs over the Internet instead of your computer's hard drive. The cloud is just a metaphor for the Internet. Cloud computing is a shared pool of computing/storage resources that can be accessed on demand and dynamically offered to the user. Cloud computing services can be accessed at any time from any place. They are offered by many companies such as Google (AWS) and Microsoft (Azure). Moreover, there are open platforms for cloud computing such as Thingspeak and Thingsboard. IoT includes heavy data (**big data**) transactions that need to be stored and analyzed. Big data concept compared to P2P or M2M is shown in Fig. 1. Cloud data is ubiquitous which means that they are available from anywhere.

2.2   Cloud Computing: Why?

The most important advantages of cloud computing can be summarized in the following points:

- **More efficient resource utilization:** You do not need a high-powered and high-priced computer to run cloud computing's web-based applications. Since applications run in the cloud, not on the desktop PC, your desktop PC does not need the processing power or hard disk space demanded by traditional desktop software. When you are using web-based applications, your PC can be less expensive, with a smaller hard disk, less memory, and more efficient processor. In fact, your PC in this scenario does not even need a CD or DVD drive, as no software programs have to be loaded and no document files need to be saved.

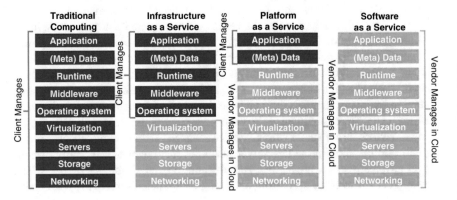

Fig. 1 Big data concept compared to P2P or M2M

- Lower cost of servers.
- Improved security.
- Reduced time-to-market.
- Virtualization.
- **Low-cost software**: Instead of purchasing expensive software applications, you can get most of what you need for free such as Google Docs suite.
- **Increased data reliability**: Unlike desktop computing, in which if a hard disk crashes and destroys all your valuable data, a computer crashing in the cloud should not affect the storage of your data. If your personal computer crashes, all your data is still out there in the cloud, still accessible.

2.3   Cloud Computing: How?

Cloud service offerings are divided into three categories: infrastructure as a service (**IaaS**), platform as a service (**PaaS**), and software as a service (**SaaS**). IaaS is responsible for managing the hardware, network, and other services. PaaS supports the OS and application platform, and SaaS supports everything. They are summarized in Table 1. A cloud system may be public or private. Public clouds can be accessed by anyone. Private clouds offer services to a set of authorized users. A hybrid cloud is a combination of both public and private clouds. In the IoT, sensors and devices communicate with one another and send data to the cloud for storage.

A cloud server is an open logical server which builds, hosts, and delivers through a cloud computing platform through the Internet. Cloud servers maintain and exhibit similar functionality as well as capabilities as a typical server. However, they are

Table 1 Cloud service models

Type	Explanation
IaaS	• Most basic cloud service model. • Cloud providers offer computers, as physical or more often as virtual machines, and other resources. • Cloud providers typically bill IaaS services on a utility computing basis, that is, cost will reflect the amount of resources allocated and consumed. • Examples of IaaS include: Google Compute Engine.
PaaS	• Cloud providers deliver a computing platform typically including operating system, programming language execution environment, database, and web server. • Application developers develop and run their software on a cloud platform without the cost and complexity of buying and managing the underlying hardware and software layers. • Examples of PaaS include: Microsoft Azure, steaming.
SaaS	• Cloud providers install and operate application software in the cloud and cloud users access the software from cloud clients. • The pricing model for SaaS applications is typically a monthly or yearly flat fee per user, so price is scalable and adjustable if users are added or removed at any point. • Examples of SaaS include: Google apps, CRM, email.

accessing remotely from a cloud service provider as open server. A cloud server is known as a virtual private server or virtual server. A cloud server is an Infrastructure as a Service (IaaS) primarily based cloud service model. Comparison of traditional computing with different service models of cloud computing is shown in Fig. 2. Figure 3 shows the cloud computing model.

There are mainly two types of **cloud servers: logical and physical**. A cloud server is a logical when it delivered through server virtualization. The physical server is distributing into two or more logical servers, each of which has a separate OS, user interface, and apps, although they share physical components from the underlying physical server, in this delivery model. At the same time, the physical cloud server is too accessing through the Internet remotely, it is not shared nor distributed. And it is commonly known as a dedicated cloud server.

Cloud computing is an example of an information technology (IT) paradigm, a model for enabling ubiquitous access to shared pools of configurable resources (such as computer servers, storage, applications and services, and networks), which can be rapidly provisioned with minimal management efforts, over the internet. Cloud computing basically allows enterprises with various computing capabilities to store and it process data either in a privately owned cloud or on a third-party server located in a data center, thus making data-accessing mechanisms more efficient as well as reliable. To achieve coherence and economy of scale similar to a utility cloud computing relies on sharing of resources.

Cloud computing allows other companies to minimize or avoid up-front IT infrastructure costs. On another hand, instead of wasting resources on computer infrastructure and their maintenance third-party clouds enable organizations to focus on their core businesses. Cloud providers coherently use a pay as you model. This might lead to unexpectedly high charges only if administrators are not familiarizing with cloud-pricing models.

Fig. 2 Comparison of traditional computing with different service models of cloud computing

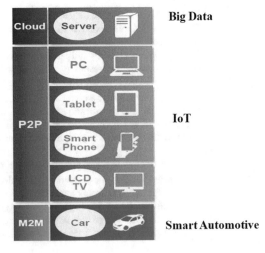

Fig. 3 Cloud computing
model

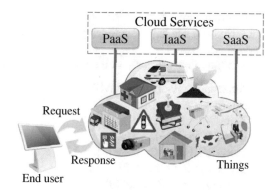

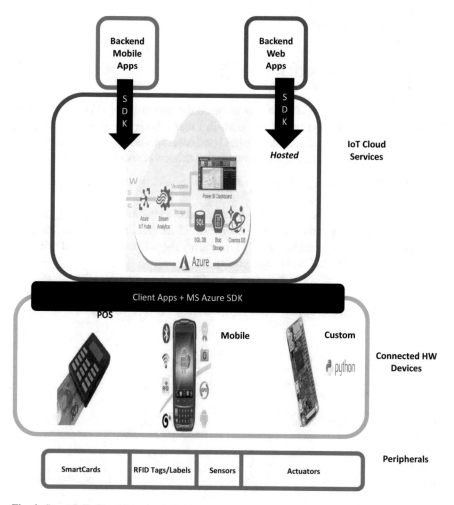

Fig. 4 Smart IoT: Cloud-based solution

Figure 4 shows a smart IoT cloud-based solution. In cloud computing, you just subscribe and use and sometimes pay for what you use based on the quality of the service (QoS).

The following are some of the top IoT cloud platforms on the market today:

- Amazon Web Services (AWS) [3].
- **Microsoft Azure** [4].
- ThingWorx IoT Platform [5].
- IBM's Watson [6].
- Cisco IoT Cloud Connect [7].
- Salesforce IoT Cloud [8].
- Oracle Integrated Cloud [9].
- GE Predix [10].
- Firebase (Google) [11].
- Parse (Facebook) [12].
- ThingSpeak [13].
- SAP [14].
- MindSphere (Siemens) [15].
- IMPACT (Nokia) [16].

MQTT, HTTP, and CoAP protocols are used for data exchange in these IoT platforms. Some criteria that have been born in mind when comparing the different platforms and how the IoT architecture has been developed in each one are explained below and comparison between different IoT platforms is provided in Table 2:

- **Flexibility**: Accepting different protocols or capability to integrate with other external systems.
- **Scalability**: When there are few devices connected to the IoT platform, the system functions correctly; the problem arises when the number of devices connected drastically increases. An IoT platform should be able to deal with the complexity associated to this increment.
- **Services**: What kind of services each platform has as strength: data store, data visualization, the device gateway—i.e., the connection between the devices that are sending data and the IoT platform—and how the different services interact.

Table 2 Comparison between different IoT platforms [17]

	Azure	AWS	IBM Watson	ThingSpeak
Protocols	HTTP, MQTT	HTTP, MQTT	HTTP, MQTT	HTTP, MQTT
Certified hardware	Raspberry Pi2, Intel	Intel, TI	Arduino UNO, Raspberry Pi, ARM	Arduino UNO, Raspberry Pi, Particle
Languages	.Net	JAVA, C	Python, JAVA, C	JAVA
Pricing	F (messages per day, number of devices)	F (messages traffic)	F (data traffic, number of devices, data storage)	Open source/free

- **Easiness of connection**: The difficulty in connecting a Thing, and specifically a Sigfox device. This is measured on the time and difficulty of creating the gateway.
- **Easiness of analysis**: How the data is stored and the easiness with which it can be accessed and transferred to visualization analysis.
- **Pricing**: Although all the platforms studied in this report have a free trial account for a period of time, and, for example, AWS can be used without paying in case the usage does not exceed a limit, it is interesting to point out the different types of payment for the same type of services used.
- **Application environment**: The possibility to develop new functions and applications according to the needs of the business strategy.

2.3.1   Google: AWS

The architecture of AWS IoT is shown in Fig. 5. The Message Broker, thing shadows, thing registry, rules engine, and the security and identity components provide the main functionality of the platform. It includes integrated data processing services, such as AWS Lambda or Amazon Kinesis, and, additionally, the IoT applications component, which enables the connection of further applications.

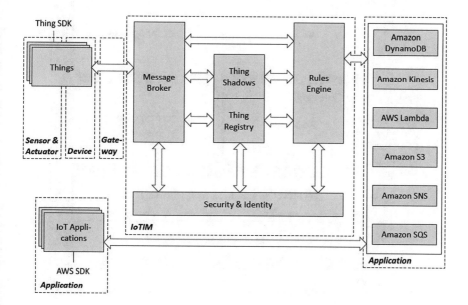

Fig. 5 AWS IoT architecture

2.3.2 IBM: Watson

The architecture of Watson IoT is shown in Fig. 6. The connect component takes care of the connection of devices to the platform and it is responsible for a corresponding message translation. Also, it provides further event handling functionality. The analytics, risk management, and information management components provide the core functionality of the platform. The Bluemix open standards based services component and the Flexible Deployment component build the basis of the platform. The IoT Industry Solutions and Third-Party Apps components enable the connection of further applications.

IBM Watson architecture allows applications to view raw data and control the behavior of devices/things. In this model, end-devices send data to the cloud using protocols. IoT gateways connect the devices that have no internet connectivity to the cloud. Real-time APIs allow detection of any changes in the data, interaction between the systems, or user interaction. The analytics unit controls the sent relevant data to the cloud with blockchain integration for secure online transactions.

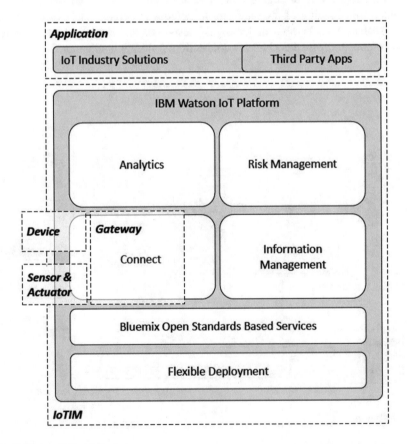

Fig. 6 Watson IoT architecture

However, this architecture has no provision for hardware abstraction layer or complex event processing. Users are not categorized and there is no provision for user registration and authentication. This architecture is also silent on different storage facilities for different applications and types of data [18].

2.3.3   Microsoft: Azure

The architecture of Azure IoT is shown in Fig. 7. The main component is the IoT Hub, where all remaining components are connected to. The core functionality of the solution is provided by the IoT Hub, the Event Processing and Insight, the device business logic, connectivity monitoring, and the application device provisioning and management components. Furthermore, the application device provisioning and management component also enables the connection of further applications [19].

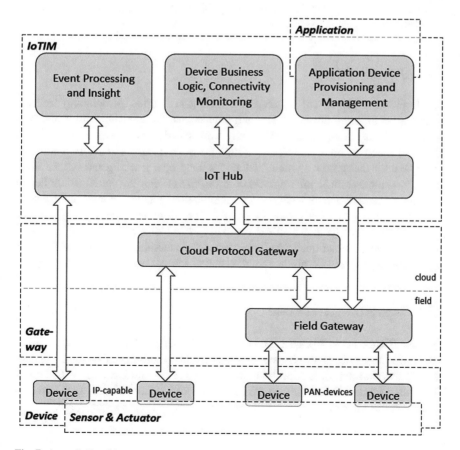

Fig. 7 Azure IoT architecture

2.3.4   ThingSpeak

ThingSpeak is an application platform for the development of IoT systems. It can help you to build the application which works upon the data collected by sensors. ThingSpeak is an open data platform for IoT application development. ThingSpeak is the perfect complement to an existing enterprise system to tap into the Internet of Things. It provides the ability to integrate your data with a variety of third-party platforms, systems, and technologies, including other leading IoT platforms such as ioBridge and Arduino. ThingSpeak channel is used to send and store data. Each channel has: eight fields that can hold any type of data, three location fields, and one status field. After creating a ThingSpeak channel, one can publish data to the channel, the data can be processed, and application can retrieve the data. ThingSpeak platform provides following functionality to support IoT system: (1) Collect: Sends sensor and device data collected from it to the cloud so that the data can be further analyzed. (2) Analyze: ThingSpeak can analyze the data received from sensors or devices and can derive the virtual representation of the data. (3) Act: Based upon the analysis, it will trigger the action to enable functioning of IoT system and application.

2.3.5   Intel IoT Platform

The Intel IoT model classifies things into various categories by including different facilities for different devices, i.e., smart devices may have business logic and analytics while the low-power devices may only consist of sensors and actuators. Things are connected to the cloud through network infrastructure. The cloud has multiple units that perform vital functions for the proper working and synchronization. The remote control unit remotely controls all the devices, monitors their health and does predictive maintenance. Communication gateway allows inter-device communication and uses protocols suitable for different types of applications. However, this architecture is silent on management of communication gateway, hardware layer abstraction. Miscellaneous storage facilities are not available in the architecture leading to failure in complex analysis as per user requirement [20].

2.3.6   Oracle Integrated Cloud

Oracle IoT Connected Worker Cloud Service helps you monitor and manage worker safety. Sensor devices worn by workers use cellular, satellite, or Bluetooth connections to transmit real-time data about the worker's status and location to Oracle IoT Connected Worker Cloud Service. You can use real-time and historic data to maintain the highest level of safety, health, and productivity for your workforce.

3   Edge/Fog Computing

Recently, there has been a move towards system architecture, namely, fog computing. What this basically does is process the time-sensitive data on the edge of the
network itself, i.e., closest to where the data is being generated so that appropriate
actions can be taken in time. Some key features of this model are: minimize latency,
conserve network bandwidth, resolve the data privacy and security issue, and
increase reliability [4, 21–23].

Edge computing provides the opportunity to reduce latency when performing
analytics in the cloud. But after data is collected on-premises, it is shared in the
public cloud. Edge computing is going to be about having some on-premises compute that works with the public cloud in a hybrid way. Microsoft Azure IoT Edge is
an example of general-purpose edge computing. Table 3 compares cloud and fog in
terms of location, size, and applications [24, 25]. Fog computing architecture is
shown in Fig. 8. The IoT is composed of three layers, cloud, fog, and device layers
as shown in Fig. 9.

Clouds are composed of servers, where each server supports applications with
computation and storage services. The device layer is composed of sensors and
actuators. Sensor data collected by sensors is delivered to servers in networks.
Sensor data is finally delivered to edge fog nodes at the fog layer. Based on the
sensor data, actions to be done by actuators are decided in the IoT. Actuators
receive actions from edge fog nodes and perform the actions on the physical environment. Fog nodes are at a layer between the device and cloud layers. Fog nodes
are interconnected with other fog nodes in networks. A fog node supports the

Table 3 Comparison between cloud and fog

	Cloud	Fog
Location	• Centralized in a small number of big data centers	• Often distributed in many locations, potentially over large geographical areas, closer to users. • Distributed fog nodes and systems can be controlled in centralized or distributed manners.
Size	• Cloud data centers are very large in size, each typically contain tens of thousands of servers.	• A fog in each location can be small (e.g., one single fog node in a manufacturing plant or onboard a vehicle) or as large as required to meet customer demands. • A large number of small fog nodes may be used to form a large fog system.
Applications	• Typically support applications that can tolerate **round-trip** delays in the order of a few seconds or longer.	• Support significantly more **time-critical** applications that require latencies below tens of milliseconds or even lower.

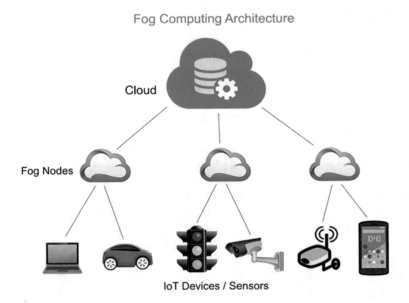

Fig. 8 Fog computing architecture [26]

routing function where messages are routed to destination nodes, i.e., routing between servers and edge nodes like network routers. More importantly, a fog node does some computation on a collection of input data sent by sensors and other fog nodes. In addition, a fog node makes a decision on what actions actuators have to do based on sensor data. Then, the edge nodes issue the actions to actuator nodes. A fog node is also equipped with storages to buffer data. Thus, data and processes are distributed to not only servers but also fog nodes in the fog computing model while centralized to servers of clouds in the cloud computing model [27].

Data from various sources and domains produced by IoT is often data stream, such as numeric data from different sensors or social media text inputs. Common data streams generally follow the Gaussian distribution over a long period. However, IoT data are produced in short time and in large quantities, presenting a variety of sporadic distributions over time. In addition, in some cases, in real time or near real time. One trend in Internet applications of Things that addresses the concept of IoT Analytics is the use of fog computing that can decentralize the processing of IoT data streams and only perform the transfer of filtered IoT data from the devices from the edge of the network to the cloud. Therefore, the association of data stream analysis with fog computing allows companies to explore in real time the data produced by IoT and thus produce business value [28].

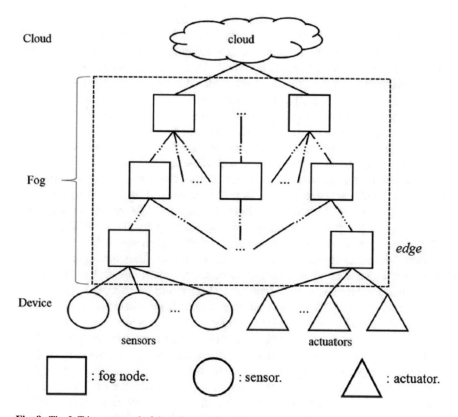

Fig. 9 The IoT is composed of three layers, cloud, fog, and device layers [27]

4 Data Analytics for Big Data: Machine Learning

The goal of IoT development is to use data, gain insight, and take action. This requires advanced analytics to visualize IoT data. Analytical techniques can be categorized into data mining, statistical analysis, **visualization,** and machine learning [29].

Artificial intelligence (AI) and machine learning (ML) models to monitor the data in the database and the incoming stream is now considered an effective solution to display and analyze data in an effective way. IoT developers don't need to be data scientists, but they do need to know how to integrate AI models into IoT databases and data streams.

In computer science, AI research is defined as the study of "intelligent agents": any device that perceives its environment and takes actions that maximize its chance of successfully achieving its goals. Colloquially, the term "artificial intelligence" is applied when a machine mimics "cognitive" functions that humans associate with other human minds, such as "learning" and "problem solving."

Machine learning (ML) has enabled breakthroughs across a variety of business and research problems, from strengthening network security to improving the accuracy of medical diagnoses. Because training and running deep learning models can be computationally demanding, Google built the Tensor Processing Unit (**TPU**), an ASIC designed from the ground up for machine learning that powers several of major products, including Translate, Photos, Search, Assistant, and Gmail. Cloud TPU empowers businesses everywhere to access this accelerator technology to speed up their machine learning workloads on Google Cloud.

TensorFlow™ is an open-source software library for high-performance numerical computation. Its flexible architecture allows easy deployment of computation across a variety of platforms (CPUs, GPUs, TPUs), and from desktops to clusters of servers to mobile and edge devices. Originally developed by researchers and engineers from the Google Brain team within Google's AI organization, it comes with strong support for machine learning and deep learning and the flexible numerical computation core is used across many other scientific domains [30].

Data Mining is also important in big data analytics. Data mining involves the discovery, collection, aggregation, transformation, matching, and processing of large datasets. Data mining is a fundamental operation incurred with the big data information system. The ultimate purpose is knowledge discovery from the data. Numerical, textual, pattern, image, and video data can be mined.

Big data is data that are too big (volume), too fast/need to be analyzed quickly (velocity), and too diverse (variety of structured and unstructured data) [31]. Examples of big data are online information (tweets, blogs), video data, voice data, and medical information. Big data is too unstructured which means that the data isn't easily put into the traditional rows and columns of conventional databases [20, 32–44]. The big data approach complements the traditional approach. In traditional approach, users determine what questions to ask and IT structures the data to answer that question. This is well suited to many common business processes, such as monitoring sales by geography, product, or channel; extract insight from customer surveys; cost and profitability analyses. In the big data approach, IT delivers a platform

Table 4 Comparison between small data and large data

Category	Small data	Big data
Definition	Data that is small enough for human comprehension	Data that is big and hard for direct human comprehension
Data source	Financial data, customer relationship management (CRM)	Sensor data, social media
Volume	$<10^{12}$	$>10^{12}$
Velocity	Does not always require an immediate response	Requires immediate response
Variety	Structured Unstructured	Structured Unstructured
Value	Predictive analysis	Reporting
Quality of data	Contains less noise	Contains more noise

that consolidates all sources of info and enables creative discover. Then the business users use the platform to explore data for idea and questions to ask. Most of the time, the data are raw data. Comparison between small data and large data is shown in Table 4 [45].

IBM Watson provides cloud-based predictive analytics for business insight, and with the cognitive computing services embedded in Watson's analytics engine, users can ask natural language questions and visualize their data patterns. Advanced analytics includes: predictive analytics, data mining, big data analytics, forecasting, text analytics, optimization, and simulation.

Hadoop is an open-source distributed processing framework that manages data processing and storage for big data applications running in clustered systems. It is at the center of a growing ecosystem of big data technologies that are primarily used to support advanced analytics initiatives, including predictive analytics, data mining, and machine learning applications. Hadoop can handle various forms of structured and unstructured data, giving users more flexibility for collecting, processing, and analyzing data than relational databases and data warehouses provide. Hadoop runs on clusters of commodity servers and can scale up to support thousands of hardware nodes and massive amounts of data [46]. Data centers in Google can be shown in Fig. 10.

Apache **Spark** is a versatile, open-source cluster computing framework with fast, in-memory analytics. It is delivered as a service on IBM Cloud [47]. **R Language** and Python can be used for big data, machine learning, and IoT [48].

4.1   IoT Analytics: Why?

- Establishes a variety of smarter environments (smarter homes, hotels, hospitals, etc.).
- Uncovers timely and actionable insights for machines and men.

Fig. 10 Data centers in Google [32]

- Enables the realization of smart objects, devices, networks, and environments.
- Leads to the production of pioneering and people-centric applications and services.
- Helps to come out with precise predictions and prescriptions.
- Facilitates process excellence and people productivity.
- Guarantees preventive maintenance of infrastructures.
- Ensures the optimized utilization of distributed assets through monitoring, measurement, and management for perfect inventory replenishment.
- Safeguards the safety and security of people and properties.
- Monitors complex environments to guarantee business performance, productivity, and resilience.

4.2   The IoT Edge Data Analytics: Real Cases

- **Manufacturing**—From creating semiconductors to the assembly of giant industrial machines, edge intelligence enhances manufacturing yields and efficiency using real-time monitoring and diagnostics, machine learning, and operations optimization. The immediacy of edge intelligence enables automated feedback loops in the manufacturing process as well as predictive maintenance for maximizing the uptime and lifespan of equipment and assembly lines.
- **Oil and gas extraction** are high-stakes technology-driven operations that depend on real-time onsite intelligence to provide proactive monitoring and protection against equipment failure and environmental damage. Because these operations are very remote and lack reliable high-speed access to centralized data centers, edge intelligence provides onsite delivery of advanced analytics and enables real-time responses required to ensure maximum production and safety.
- **Mining** faces extreme environmental conditions in very remote locations with little or no access to the Internet. As a result, mining operations are relying more and more on edge intelligence for real-time, onsite monitoring and diagnostics, alarm management, and predictive maintenance to maximize safety, operational efficiency, and to minimize costs and downtime.
- **Transportation**—As part of the rise in the Industrial Internet, trains and tracks, buses, aircraft, and ships are being equipped with a new generation of instruments and sensors generating petabytes of data that will require additional intelligence for analysis and real-time response. Edge intelligence can process this data locally to enable real-time asset monitoring and management to minimize operational risk and downtime. It can also be used to monitor and control engine idle times to reduce emissions, conserve fuel, and maximize profits.
- **Power and Water**—The unexpected failure of an electrical power plant can create substantial disruption to the downstream power grid. The same holds true when water distribution equipment and pumps fail without warning. To avoid this, edge intelligence enables the proactive benefits of predictive maintenance and real-time responsiveness. It also enables ingestion and analysis of sensor

data closer to the source rather than the cloud to reduce latency and bandwidth costs.

- **Renewable Energy**—New solar, wind, and hydro are very promising sources of clean energy. However, constantly changing weather conditions present major challenges for both predicting and delivering a reliable supply of electricity to the power grid. Edge intelligence enables real-time adjustments to maximize power generation as well as advanced analytics for accurate energy forecasting and delivery.
- **Healthcare**—In the healthcare industry, new diagnostic equipment, patient monitoring tools, and operational technologies are delivering unprecedented levels of patient care but also huge amounts of highly sensitive patient data. By processing and analyzing more data at the source, medical facilities can optimize supply chain operations and enhance patient services and privacy at a much lower cost.
- **Retail**—To compete with online shopping, retailers must lower costs while creating enhanced customer experiences and levels of service that online stores cannot provide. Edge intelligence can enrich the user experience by delivering real-time omni-channel personalization and supply chain optimization. It also enables newer technologies such as facial recognition to deliver even higher levels of personalization and security.
- **Smart Buildings**—Among the many benefits of smart building technology are lower energy consumption, better security, increased occupant comfort and safety, and better utilization of building assets and services. Rather than sending massive amounts of building data to the cloud for analysis, smart buildings can use edge intelligence for more responsive automation while reducing bandwidth costs and latency.
- **Smart Cities**—Integrating data from a diverse collection of municipal systems (e.g., street lighting, traffic information, parking, public safety) for interactive management and community access is a common vision for smart city initiatives. However, the sheer amount of data generated requires too much bandwidth and processing for cloud-based systems. Edge intelligence provides a more effective solution that distributes data processing and analytics to the edges where sensors and data sources are located.
- **Connected Vehicles**—Connected vehicle technology adds an entirely new dimension to transportation by extending vehicle operations and controls beyond the driver to include external networks and systems. Edge intelligence and fog computing will enable distributed roadside services such as traffic regulation, vehicle speed management, toll collection, and parking assistance.
- **Quality Management**—In this era of intense competition, maintaining highest level of product quality is key to beating competition. Assembly lines produce thousands of units every day. They go through robust testing process before being shipped out. However, there are instances when defects escape tests. Factories learn about these defects only after customers log complaints. There have been multiple instances when products have been recalled due to defects. With IoT, plants can collect large volume of manufacturing and testing data in

real time. It will then be worked on by advanced analytics module in the IoT platform to match the pattern with the ones collected from defective products. Whenever the patterns are similar or same, the lot in question can be sent back to the factory to undergo further testing. This will help preempt the defects and fix them before customers log complaint. Even in cases when defects are analyzed post customer complaint, it's challenging in the existing analytics framework to nail down the exact lot of the materials used in them. This is because the volume of data is so large, traditional analytics platform may not be able to effectively handle them. IoT platforms specialize in analyzing large volume of data aka "big data," hence they can provide more accurate analysis as to where the defect originated [49, 50].

- **Optimization**—Plants consume many resources such as raw materials, energy, and water. There are two ways IoT can help optimize. First, in a process manufacturing environment, a given amount of chemical is used to produce multiple lots of finished products. Amount consumed is not uniform all the time. IoT can be of help in this case, they can collect real-time consumption data directly from measuring instruments in equipment and analyze them over hundreds of lots. They can then point to possible anomalies which might be responsible for overconsumption in some cases. Fixing this can help save so much money for the factories. Similarly, IoT platforms can collect and analyze energy consumption data of equipments, air-conditioning system, lighting systems etc. and perform complex what-if analysis and suggest an optimum usage pattern [51].

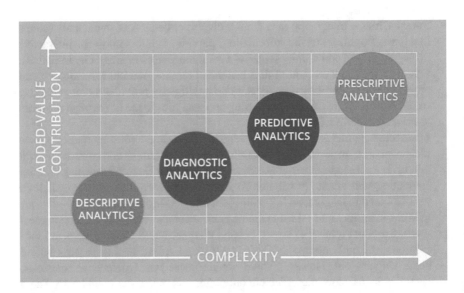

Fig. 11 Data analytics types. Machine learning in IoT can be used for forecasting of information patterns

4.3   IoT Data Analytics: Types

There are four types of data analytics as depicted in Fig. 11. The classification depends on what you want to do with the data, whether you need to generate reports, predict information or do recommendations.

- **Descriptive Analytics**: find an answer to question "what happened?", generate reports or visualize the data.
- **Diagnostic Analytics**: find an answer to question "why did it happen?", find correlations.
- **Predictive Analytics**: find an answer to question "what will happen?", it uses machine learning.
- **Prescriptive Analytics**: find an answer to question "how can we make it happen?", it uses more complicated machine learning algorithms to find recommendation.

5   Conclusions

This chapter presents the integration of cloud computing and IoT, including the data analysis. Moreover, it gives insights into edge/fog computing. Due to the massive datasets required, machine learning is a requirement for IoT to provide value. Moreover, we present the importance of data analytics from sensors to decisions. IoT holds a lot of promise in manufacturing area. Machine learning in IoT can be used for forecasting of information patterns.

References

1. What is an IoT application platform? | Zatar. Zatar.com, 2017. [Online]. Retrieved August 13, 2017, from http://www.zatar.com/blog/what-is-an-iot-application-platform.
2. Botta, A., De Donato, W., Persico, V., & Pescapé, A. (2016). Integration of cloud computing and internet of things: A survey. *Future Generation Computer Systems, 56,* 684.
3. Retrieved from http://aws.amazon.com
4. Mouradian, C., Naboulsi, D., Yangui, S., Glitho, R. H., Morrow, M. J., & Polakos, P. A. (2018). A comprehensive survey on fog Computing: State-ofthe- art and research challenges. *IEEE Communications Surveys and Tutorials, 20*(1), 416–464.
5. Retrieved from https://www.ptc.com/en/resources/iot/product-brief/thingworx-platform
6. Retrieved from https://www.ibm.com/cloud-computing/
7. Retrieved from https://www.cisco.com/
8. Retrieved from https://www.salesforce.com/eu/products/iot-cloud/overview/
9. Retrieved from https://www.oracle.com.
10. Retrieved from https://www.ge.com/digital/iiot-platform.
11. Retrieved from https://cloud.google.com/
12. Retrieved from www.facebook.com

13. Retrieved from www.thinkspeak.com
14. Retrieved from www.sap.com
15. Retrieved from www.siemens.com
16. Retrieved from https://networks.nokia.com/solutions/iot-platform.
17. Ray, P. P. (2016). A survey of IoT cloud platforms. *Future Computing and Informatics Journal, 1*, 35e46.
18. Webcast: IoT use-cases with IBM Watson IOT Platform. [Online]. Retrieved September 10, 2017, from https://marionoioso.com/2016/11/05/webcast-iot-use-cases-with-ibm-watson-iot-platform/. IBM Watson IoT Architecture.
19. Aazam, M., Khan, I., Alsaffar, A.A., Huh, E.N. (2014). Cloud of things: Integrating internet of things and cloud Computing and the issues involved. *International Bhurban Conference on Applied Sciences and Technology*. IEEE.
20. George, G., Haas, M. R., & Pentland, A. (2014). Big data and management. *Academy of Management Journal, 57*(2), 321–326.
21. Pan, J., & McElhannon, J. (Feb. 2018). Future edge cloud and edge Computing for internet of things applications. *IEEE Internet of Things Journal, 5*(1), 439–449.
22. Bilal, K., Khalid, O., Erbad, A., & Khan, S. U. (Jan. 2018). Potentials, trends, and prospects in edge technologies: Fog, cloudlet, Mobile edge, and micro data centers. *Computer Networks, 130*, 94–120.
23. Atlam, H. F., Walters, R. J., & Wills, G. B. (2018). Fog Computing and the internet of things: A review. *Big Data and Cognitive Computing, 2*(10), 1–18.
24. Botta, A., De Donato, W., Persico, V., et al. (2016). Integration of cloud computing and internet of things[J]. *Future Generation Computer Systems, 56*(C), 684–700.
25. Chiang, M., & Zhang, T. (2016). Fog and IoT: An overview of research opportunities[J]. *IEEE Internet of Things Journal, 3*, 854–864.
26. Mahmud, R., Buyya, R. (2018). Fog Computing: A taxonomy, survey and future directions. *Internet of Everything* (pp. 103–130), Springer.
27. Oma, R., Nakamura, S., Duolikun, D., Enokido, T., Takizawa, M. (2018). An energy-efficient model for fog computing in the internet of things (IoT). *Internet of Things*, 1-2, 14. Elsevier.
28. Leandro Andrade. SOFT-IoT Platform in Fog of Things. WebMedia '18, October 16–19, 2018, Salvador-BA, Brazil.
29. Mohamed, K. S. (2018). *Machine learning for model order reduction*. Springer.
30. Retrieved from https://www.tensorflow.org/
31. Madden, S. (2012). From databases to big data. *IEEE Internet Computing, 16*(2012), 4–6.
32. Agarwal, R., & Dhar, V. (2014). Editorial—Big data, data science, and analytics: The opportunity and challenge for IS research. *Information Systems Research, 25*(3), 443–448.
33. Lim, E. P., Chen, H., & Chen, G. (2013). Business intelligence and analytics: Research directions. *ACM Transactions on Management Information Systems (TMIS), 3*(4), 17.
34. Dhar, V. (2013). Data science and prediction. *Communications of the ACM, 56*(12), 64–73.
35. Raghupathi, W., & Raghupathi, V. (2014). Big data analytics in healthcare: Promise and potential. *Health Information Science and Systems, 2*(1), 3.
36. Khan, Z., Anjum, A., Soomro, K., & Muhammad, T. (2015). Towards cloud based big data analytics for smart future cities. *Journal of Cloud Computing: Advances, Systems and Applications, 4*.
37. McAfee, A., & Brynjolfsson, E. (2012). *Big data: The management revolution*. Harvard business review.
38. Davenport, T. H., & Patil, D. J. (2012). *Data scientist*. Harvard business review.
39. Provost, F., & Fawcett, T. (2013). Data science and its relationship to big data and data-driven decision making. *Big Data, 1*(1), 51–59.
40. Yeoh, W., & Koronios, A. (2010). Critical success factors for business intelligence systems. *Journal of Computer Information Systems, 50*(3), 23.
41. Davenport, T. H. (2012). *Enterprise analytics: Optimize performance, process, and decisions through big data*. FT Press.

42. Minelli, M., Chambers, M., & Dhiraj, A. (2013). *Big Data, Big Analytics: Emerging business intelligence and analytic trends for Today's businesses*. Hoboken, NJ: Wiley.
43. Mayer-Schonberger, V., & Cukier, K. (2013). *Big Data: A revolution that will transform how we live, work, and think*. Eamon Dolan/Houghton Mifflin Harcourt.
44. Dean, J. (2014). *Big data, data mining, and machine learning: Value creation for business leaders and practitioners*. Hoboken, NJ: Wiley.
45. Agrawal, D.; Bernstein, P.; Bertino, E.; Davidson, S.; Dayal, U.; Franklin, M.; Widom, J. (2012). *Challenges and opportunities with big data: A white paper prepared for the Computing community consortium committee of the Computing Research Association*. Retrieved November 13, 2018, from http://cra.org/ccc/docs/init/bigdatawhitepaper.pdf.
46. Retrieved from https://software.intel.com/sites/default/files/article/402274/etl-big-data-with-hadoop.pdf
47. Retrieved from https://www.ibm.com/cloud/spark
48. Retrieved from https://iotmakerblog.wordpress.com/2016/07/03/r-tutorials/
49. Retrieved from https://services.mesa.org/ResourceLibrary/ShowResource/7e1651ec-41c6-4d86-9c1e-6f298fb54756.
50. Retrieved from https://services.mesa.org/ResourceLibrary/ShowResource/ec1ddaf4-5aa6-4710-9027-47f4ba3ab43f.
51. Retrieved from http://dev.expotrademe.com/FMTS2017-PPT-PDFs/.

IoT Application Layer: Case Studies and Real Applications

1 Introduction

Internet of Things (IoT) is a growing industry. Analysts predict that (IoT) products and services will grow exponentially in next years. It is a confluence of different sectors: embedded systems, communication systems, sensors/actuators, WWW, and mobile applications. Use Internet of Things Technology to solve all problems in different life sectors: healthcare, museums, libraries, inventory management, advertisement, real-estate identification, food tracking, maintenance, radiation/pollution monitoring, and security [1].

However, the Internet of Things is still maturing, in particular due to a number of factors, which limit the full exploitation of the IoT:

- No clear approach for the utilization of unique identifiers and numbering spaces for various kinds of persistent and volatile objects at a global scale.
- No standard IoT reference architecture.
- Less rapid advance in semantic interoperability for exchanging sensor information in heterogeneous environments.
- Difficulties in developing a clear approach for enabling innovation, trust, and ownership of data in the IoT while at the same time respecting security and privacy in a complex environment. Difficulties in developing business which embraces the full potential of the Internet of Things. In our proposal, we believe that presenting simple yet clear services which can be mixed together to create complex scenarios to interact and react with Things around us.
- Government regulations on usage of GPS in (Geolocation) and restrictions on communication systems that interfere with police and military sector. In our proposal, we can easily switch between WIFI and 3G/GPRS. We believe that these regulations will be relaxed in future as situations get better and technology evolution imposes the change.

© Springer Nature Switzerland AG 2019
K. S. Mohamed, *The Era of Internet of Things*,
https://doi.org/10.1007/978-3-030-18133-8_5

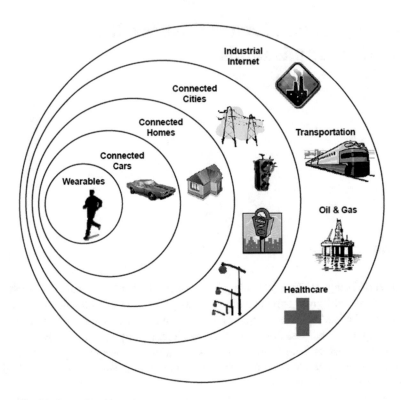

Fig. 1 IoT landscape/ecosystem

Different applications will be discussed in this chapter. IoT landscape is shown in Fig. 1. IoT applications can be classified into two major applications: consumer applications and business/enterprise applications.

2   IoT Case Studies

Internet of things' application layer holds the responsibility for providing services and defines the set of messages' protocols that are passing at this level. There must be some data processing environment for analyzing the data fetched from the devices (sensors, controllers, etc.) and making this data usable. Thus, this usability can be through direct applications with easy graphical user interface (GUI) for individual users, like mobile applications for simple IoT applications, or for massive projects that host global users; clouds can be used to analyze, sort out and store the data; and websites can be used as an interface [2–14].

The application layer is also concerned with providing a virtual service layer that is responsible for data transport, security, and service discovery and device management on a high level of abstraction, independent of communication technologies in the lower layers. This ensures the right connectivity between devices and various IoT applications to realize horizontally integrated IoT for specific applications. This virtual service layer provides information collected from objects and the performance of the actuators. For instance, while data from a temperature sensor for home automation are provided, it should also describe if it is the indoor temperature of a room, or a fridge, etc. IoT potential allows it to customize any kind of applications. Applications of our solutions can be highlighted in the following examples [15–21].

2.1 Hospital Model/e-Health

The Internet of Things (IoT) has enabled remote sensing and communication with various devices. In the area of healthcare, IoT has far benefits in monitoring and alerting patients. IoT healthcare is applicable in many medical instruments such as ECG monitors, glucose level sensing, and oxygen concentration detection. This model is used at healthcare facilities to identify medicine, medical equipment, and patients or new born babies. Each will have a tag or label and a connected device will read the tag and communicate tag info with IoT cloud. Backend application will identify the object, and communicate with a host database to retrieve all information about identified object. e-Health example is shown in Fig. 2.

IoT has the potential to transform healthcare. Several successful IoT applications already exist in the healthcare sector, covering patient monitoring and treatment and

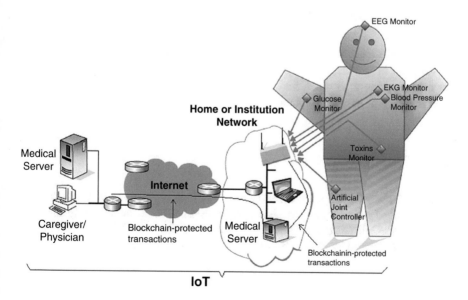

Fig. 2   e-Health example [22]

hospital management. Remote monitoring of patients is a key focus area for high investment because of the expected improved outcomes. IoT has the potential to help patients and their doctors be more effective at managing chronic diseases, which is a growing imperative across the healthcare system.

2.2 Museum Model

This model is used at museums, exhibitions, and sightseeing areas as a guidance system for tourists to identify the location and physical objects at these spots. This model is capable of displaying information to user or providing explanation in sound through a headphone and locating the current position on a digital map of the sightseeing place.

2.3 Inventory Model

This model is used at factories and supermarkets to automatically count, identify box contents and location. Data analytics on inventory database will help answer the questions: "where can I distribute my products more?" and "when can I increase my production rate?" and "where are my products distributed?".

2.4 Advertising Model

This model is used at shopping malls where each shop will supply a vide blog for advertising its products. Each product or a shop poster will have a tag or label and customers using their connected devices will be capable to display advertisement previews and promotions on their backend application.

2.5 Food Tracing Model

This model is used to track the history record of food products by adding a tag corresponding to each production phase. In the event of incidents affecting the food quality, traceability is designed to enable us identify the cause quickly and confirm the scope of impact by tracing how the food was transported.

2.6   Residence Model

This model is used to identify locations like road address, building address, and public stations. New buildings can have a trial video on its construction phases, prices, plans and constructors contacts.

2.7   Maintenance Model

This model is used to track the maintenance time of facilities, by spreading tags on every object in a given place that require maintenance. A manager can then identify each object and track its maintenance schedule. A useful model can be applied to public facilities, e.g., metro or companies.

2.8   Fire Alarm Model

This model is used to sense smoke, temperature in different connected locations and send to IoT cloud. Backend application will monitor and tweet/alarm user on any change in risk of fire.

2.9   Attendance Model

This model is used instead of time cards, to monitor the absence and localization of employees, visitors, students within a company, sporting club, or academic institute. Backend application can gather reports, link with an HR system or do analytics on attendance and absence.

2.10   Access Control Model

This model is used to restrict the access to locations (doors/gates). It records the time and identifies the person accessing the location and transfers this information to IoT cloud to localize that person and achieve his access permission.

2.11   Library Model

This model is used at libraries where each book will have a tag or label to identify the book. The readers will have a connected device to get information about books and its availability to user from IoT cloud. A librarian can use a backend application connected to IoT cloud to manage borrowing books.

2.12   Cashless Payment Model

This model includes a smart card (for medical examination, meals, and sporting club facilities reservations, transportation, or parks services) and a "Point of Sale (PoS)"—connected device to an IoT cloud. Backend application can identify a person granted a given service and deduct from his balance. The cashless payment model is shown in Fig. 3.

2.13   Connected Animal Model

This model is used at farms to monitor the healthcare of live animals, identify and automatically count them. It can sense temperature and pH of a connected aquarium for healthy fish. Backend application can visualize data, and do analytics on gathered data [23].

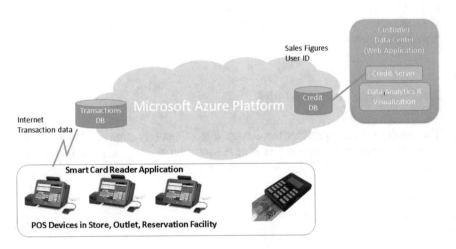

Fig. 3 The cashless payment model

2.14   Connected Plant Model

This model is used at farms to monitor the healthcare of plants (temperature, humidity, and air quality). Backend application can visualize data, and do analytics on gathered data.

2.15   Connected Police Model

This model is used for military or security companies to monitor health and location of connected security men. Assist to identify a watch list person. Stream in real time from a connected camera upon engagement in a riot.

2.16   E-Commerce Model/Smart Supply Chain

E-commerce is the buying and selling of products through the use of internet. The buyer orders a product online and pays the money electronically. The seller and buyer are not required to come face to face and shopping can be done smoothly. After the order is confirmed, the seller transports the shipment and the tracking details are given to the customer. Then the customer receives the product. These days e-commerce has become a very common method of business these days [24]. Moreover, IoT can improve supply chain management. Supply chain management is defined as a strategic way, which by means of integration the intermediates of supply chain and coordinating cooperation of all parties involved in exchange of information, materials and money, maximizes benefits and performance of entire supply chain [3].

2.17   Smart Cities

IoT can be used to build smart cities [25, 26]. This includes smart transportation, smart hospitals, smart schools, smart traffic control, smart banking, smart vehicles, smart parking, smart environment (waste management, population management, weather monitoring), and smart police. Smart refers to smart services that will be offered by the cities to the citizens [27]. Smart city uses information technologies and IoT to improve the quality of life, efficiency, and economic competitiveness of the city while ensuring a sustainable growth and protecting the environment [28, 29]. IoT can be very useful for resource management in the context of smart cities.

Fig. 4 Autonomous car
architecture

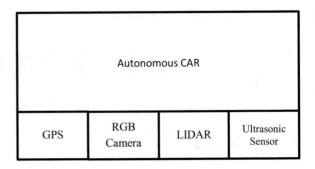

2.18   Smart Vehicle

Figure 4 shows an IoT-enabled autonomous and smart car, where the car is equipped with many sensors, different networking and communication devices. Autonomous cars are complex, inter- and intra-dependent systems that include multiple levels of hardware and software. The hardware shapes the methods of input and output. Sensors gather information from GPS/Inertial Measurement Units (IMUs), cameras, LiDAR, and radar. Vehicle-to-vehicle (V2V) and vehicle-to-infrastructure (V2I) components enable vehicles to capture data from and communicate with each other as well as signal lights, signage, roads, etc. The final pieces of hardware are the actuators, which enable control and actually move the machine.

2.19   Smart Homes

Smart home is one of the applications of IoT that continues to grow rapidly. It contains smart devices that are connected with applications and gateways. Applications of smart homes include smart lights, smart locks, smart TVs, and smart washing alarm system. Smart home applications have gained a huge attention recently. By using IoT technology, home appliances are able to communicate with each other so the user can have access to all devices in the home anytime and anywhere. You can control and monitor your home from your Android Smartphone or Tablet. Smart homes services include [30, 31]:

- Security and safety (locks, camera, smoke detectors).
- Health monitoring.
- Resource savings (lighting, heating).
- Remote monitoring.

  Big internet companies try to build platforms for smart home such as:

- Apple: HomeKit.
- Google: Nest.
- Amazon: Echo.

  Moreover, there are many open-source home automation solutions such as OPENHAB [32]. The general framework for Arduino-based home automation is shown in Fig. 5. Home lighting application is also shown in Fig. 6.

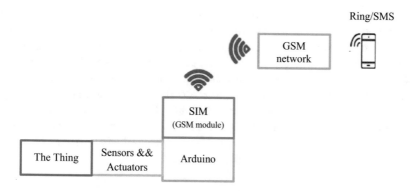

Fig. 5 The general framework for Arduino-based home automation

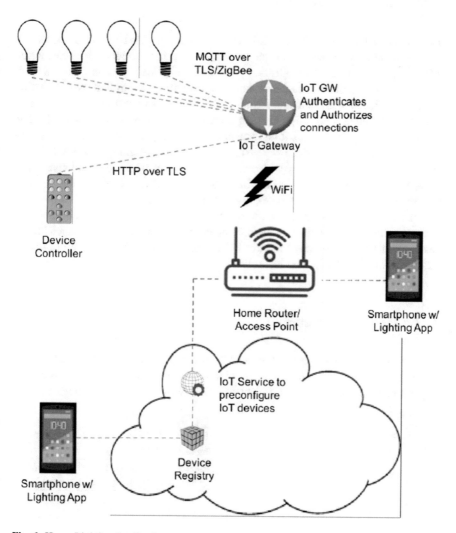

Fig. 6 Home Lighting Application

The entire architecture of smart homes needs some se requirements against existing security issues [33]. That's why, there's huge need for strong authentication mechanism to prevent from attacks. The security attacks in smart home can be categorized into two categories: passive attacks and active attacks. The passive attacks such as eavesdropping aim to listen without modifying the data as the attacker obtains information from the system, monitors the system, transmits the messages and does not modify them but he learns something from it. Generally, these types of attacks are undetectable. The active attacks aim to modify the data or the messages, breaking into the network equipment. Active attacks include denial of service (DoS), message modification, and password cracking [34].

2.20 Smart Factories/IIoT

The industrial internet of things (IIoT) is an application of IoT in industries. IIoT is considered as an intersection between industries and IoT. The industrial revolution 4.0 is happening today through the use of cyber-physical systems. It means that physical systems such as machines and robotics will be controlled by automation systems equipped with machine learning algorithms. Minimal input from human operators will be needed [35]. Oil and gas are one of the reasons IIoT is so big. The aim is to improve data acquisition (sensors and robots), processing and interpretation, reduce downtime, reduce site visits, remote monitoring, increase productivity and reduce accident frequency with real-time monitoring of assets, predictive maintenance [36]. Predictive maintenance in the Internet of Things (IoT) era can be summarized as a maintenance methodology that brings together the power of machine learning and streaming sensor data to maintain machines before they fail, optimize resources, and thereby reduce unplanned downtime. Predictive maintenance identifies manufacturing equipment failures before they happen. The evolution of industrial revolution is summarized in Table 1.

2.21 Smart Grid/Energy

It is new challenges with renewable sources. Smart grid involves distributed generation and information networks, remote monitoring of failures [37, 38]. Smart grid technologies all contribute to efficient IoT energy management solutions that are

Table 1 Evolution of industrial revolution		
	Industry 1.0	Steam power and mechanics
	Industry 2.0	Electricity
	Industry 3.0	Computers and automation
	Industry 4.0	Cyber-physical systems

currently lacking in the existing framework. What makes the IoT smart grid better is two-way communication between connected devices and hardware that can sense and respond to user demands. These technologies mean that a smart grid is more resilient and less costly than the current power infrastructure [39]. Smart grid is an excellent solution to optimize the energy consumption.

2.22   Smart Environment

Advances in many technical areas are making the IoT and smart environments possible, including multiple communication solutions for IoT devices. While all smart environments collect, process, and act upon information, different specific smart environments do so at different scales [40–42]. Here are some applications:

- **Forest Fire Detection**: Monitoring of combustion gases and preemptive fire conditions to define alert zones.
- **Air Pollution**: Control of CO_2 emissions of factories, pollution emitted by cars, and toxic gases generated in farms. Air Quality (AQ) is a very topical issue for many cities and has a direct impact on citizen health [43].
- **Landslide and Avalanche Prevention**: Monitoring of soil moisture, vibrations, and earth density to detect dangerous patterns in land conditions.
- **Earthquake Early Detection**: Distributed control in specific places of tremors.
- **Water Quality:** Study of water suitability in rivers and the sea for fauna and eligibility for drinkable use.
- **Water Leakages:** Detection of liquid presence outside tanks and pressure variations along pipes.
- **River Floods:** Monitoring of water level variations in rivers, dams, and reservoirs [44].

2.23   Smart Agriculture

Smart farming is becoming more cost effective. Advances in standards and protocols driven by interest in the internet of things are allowing for more choice in design. The increased availability enables for designs that are applicable for the most challenging of applications, that of subsistence agriculture in the developing world. The single most serious effect of climate change to date in the global south is that of water supply. The well-defined wet and dry seasons are no longer well defined. Solar power together with microelectronics on boards and low-power short-range RF links can provide sensor fusion and actuator control at moderate cost. Near real time monitoring and evaluation is more the challenge [45].

The aim of smart agriculture:

- Producing more by optimizing crop yields.
- Cheaper production by using fewer chemicals, fertilizers.
- Better use of natural resources: water, etc.
- Improve rural area connectivity.
- Analyze collected data to make the right decision.
- Direct command of watering and fertilizer distribution.
- Plant leaf diseases detection and auto-medicine [46].
- Yield and soil monitoring.
- Precise irrigation.

2.24 Smart Roads/Streets/Traffic: Smart Lane Divider

Road divider is generically used for dividing the road for ongoing and incoming traffic. This helps keeping the flow of traffic. Generally, there is equal number of lanes for both ongoing and incoming traffic. For example, in any city, there is industrial area or shopping area where the traffic generally flows in one direction in the morning or evening. The other side of road divider is mostly either empty or underutilized. This is true for peak morning and evening hours [47]. These results in loss of time for the car owners, traffic jams as well as underutilization of available resources. The idea is to formulate a mechanism of automated movable road divider that can shift lanes, so that we can have more number of lanes in the direction of the rush [48].

2.25 Chabot

Chatbots are artificial intelligence systems that interact with humans through chat interfaces via messaging, text, or speech. A chatbot can respond to certain questions or give recommendations on different topics in a real-time manner. In an IoT environment, chatbots can function as an interface to make sense of all the data and also make it more accessible. Facebook Messenger's chatbot SDK provides companies a platform where they can integrate their own service within the accessibility of the Messenger app [49].

2.26 Smart Education

In education, mobile-enabled solutions will tailor the learning process to each student's needs, improving overall proficiency levels, while linking virtual and physical classrooms to make learning more convenient and accessible. The students don't have to carry heavy books every day, also results are sent to the school and the reports are updated in real time. Besides, from the same smart device, a student can connect to classmates and teachers. Mobile education solutions have already been shown to improve learners' proficiency rates and reduce dropout rates, and have the potential to enable, by 2017, the education of up to 180 million additional students in developing countries who will be able to stay in school due to this system.

Nowadays, many universities and schools have launched the initiative of smart classrooms. However, there is no unique definition for smart classrooms. They are generally normal classrooms enhanced with smart technologies that facilitate the teaching and the learning process. They include a set of hardware devices and software platforms. In a smart classroom, teachers do not spend time to record attendance, and it is done automatically. The smart chairs system facilitates classroom management and interaction, and it tracks attendance [50–52].

2.27 Smart Club

The smart club framework can be viewed in Fig. 7. It contains the following tasks:

- Time Attendance.
- Access Control: Entrance/Exit to club (RFID enabled smart cards).
- Members Management.
- Cashless Payment (Smart Cards): Reservation of facilities or Payments at outlets.

Fig. 7 The smart club framework

2.28   Wearables

Major technology companies such as Apple, Google, Samsung, and Intel invest heavily in wearables, with non-tech giants like Nike, Under Armour, and Adidas. Most famous wearables are summarized in Table 2. Some wearables examples are also shown in Fig. 8. Wearable functionalities can be for: fitness and tracking, contactless payment, health as already some devices available for seniors and babies (alerting and sleep monitors), identification, authentication, and localization. Figure 9 shows an example of how wearable works where dedicated application runs on the smartphone and forwards the data to a cloud platform.

2.29   Smart Tourism

Tourism is one of the major components of economy growth of many countries in world. Due to the lack of coordinated services, tourism has suffered a lot. In [55], an IoT framework called iTour has been presented which was implemented in Java. In iTour, the smart citizens can participate in tourism development. It enables the city administrator accountable towards the cooperation and coordination of tourist daily life in a city. It can also use data mining techniques to prepare the city for future tourism. The effectiveness of iTour is being proven through in-depth evaluation in a smart city by involving all stakeholders.

2.30   3D Printing

Sooner or later, silicon circuits will be replaced by plastic circuits. And 3D printing technologies will dominate such plastic printed electronics and circuits in IoT. Plastic printed transistors will become building blocks of wearable electronics and other IoT networks and the whole thing will be much cheaper than silicon devices, semiconductors, and circuits [56].

Table 2 Wearables examples

Wearable	Manufacturer
Google glass (smart glasses)	Google
Jawbone (activity monitor)	MotionX
Baby status monitoring	Mimo
Apple watch [53]	Apple

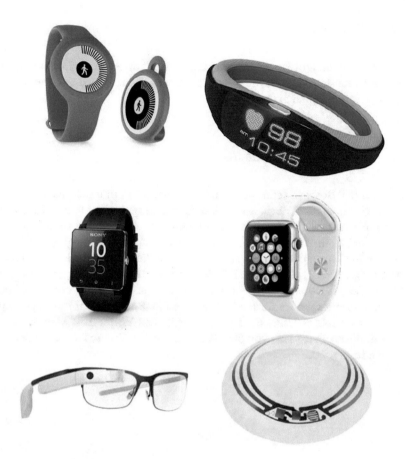

Fig. 8 Examples of IoT wearables

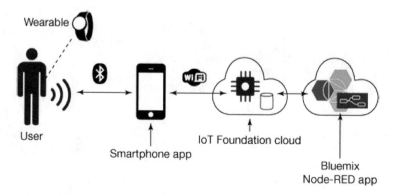

Fig. 9 How wearable works [54]

2.31 3D Scanning

3D scanning is the process of analyzing a real-world object or environment to collect data on its shape and possibly its appearance. The collected data can then be used to construct digital 3D models [57].

2.32 ARP and CRM Systems

The Odoo IoT Box is a small box, with a Raspberry Pi in it that makes it easier for you to connect IoT devices to Odoo. Connecting devices to an ERP system is often difficult. Many of the devices in question will not have an internet connection, and even if they do, they will not automatically integrate with your ERP. The IoT Box solves this by serving as a middle point between your devices and your Odoo ERP system. The IoT Box connects to Odoo through an Ethernet cable or Wi-Fi. Odoo detects the boxes, and you can easily configure them through your browser. As for the devices, the IoT Box allows for multiple input connections, such as USB, Bluetooth, HDMI, and Wi-Fi. This way, whatever connection the device has, you can likely connect it. The IoT Box architecture is shown in Fig. 10.

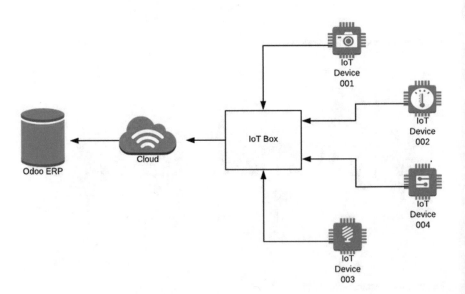

Fig. 10 Odoo IoT Box example [58]

3 Conclusions

This chapter presents a review of major technologies of IoT-based applications. IoT applications are nearly limitless. IoT can be used into all the domains including medical, manufacturing, industrial, transportation, education, governance, mining, etc. Moreover, we introduce a comparative analysis of various models used by industry for developing IoT applications. IoT will have a great impact on the business in the future.

References

1. F. (2017). Smart Home Internet of Things| Xively by LogMeIn. Xively.com. [Online]. Retrieved July 18, 2017, from https://www.xively.com/smart-home-internet-of-things.
2. Miorandi, D., Sicari, S., De Pellegrini, F., & Chlamtac, I. (2012). Internet of things: Vision, applications and research challenges. *Ad Hoc Networks, 10*(7), 1570–8705.
3. Zhu, J., & Fu, B. B. (2015). Research on supply chain simulation system based on internet of things. *Advances in Internet of Things, 5*, 1–6.
4. Kurde, A., & Kulkarni, V. S. (2016). IOT Based Smart Power Metering. *International Journal of Scientific and Research Publications., 6*(9), 411–415.
5. Kumar, A., & Tiwari, N. (2015). Energy efficient smart home automation system. *International Journal of Scientific Engineering and Research, 3*(1), 9–11.
6. Retrieved from https://www.iot-now.com/2017/03/09/59388-iot-applications-autonomous-vehicles-smarter-cars-cellular-iot-vehicle-telematics/.
7. Khanna, A., & Anand, R. IoT based smart parking system. *International Conference on Internet of Things and Applications (IOTA).* doi:https://doi.org/10.1109/IOTA.2016.7562735.
8. Chiuchisan, I. & Dimian, M. (2015). Internet of things for eHealth: An approach to medical applications. *International Workshop on Computational Intelligence for Multimedia Understanding (IWCIM).*
9. Sun, C. (2012). Application of RFID Technology for Logistics on internet of things. *AASRI Procedia, 1*, 106–111.
10. Da Xu, L., He, W., & Li, S. (Nov. 2014). Internet of things in industries: A survey. *IEEE Transactions on Industrial Informatics, 10*(4), 2233.
11. Gaonkar, S., & Meghashree, A. C. (2017). Emergency Tracking system for women using body Sensors via Wrist watches using Internet of Things (IOT). *International Research Journal of Engineering and Technology (IRJET), 4*(8), 55–58.
12. Stewart, J., Stewart, R., & Kennedy, S. (2017). Internet of things—Propagation modeling for precision agriculture applications. *Wireless Telecommunications Symposium (WTS).* https://doi.org/10.1109/WTS.2017.7943528.
13. Merchant, H. K., & Ahire, D. D. (2017). Industrial automation using IoT with raspberry pi. *International Journal of Computer Applications., 168*, 44–48.
14. Ray, P. P. (2016). Internet of robotic things: Concept, technologies, and challenges. *IEEE Access, 4*, 9489. https://doi.org/10.1109/ACCESS.2017.2647747.
15. Mehta, Y., & Mounika, J. (2017). 5 internet of things (IoT) trends in 2017-19— IoT worm. *IoT Worm.* [Online]. Retrieved August 13, 2017, from http://iotworm.com/5-internet-things-iot-trends-2017-19/.
16. Cisco. (2017). *Smart Traffic Management with Real Time Data Analysis.* [Online]. Retrieved July 18, 2017, from http://www.cisco.com/c/en_in/about/knowledge-network/smart-traffic.html.

17. Mounika, J., Prince, A., Mehta, Y., Nivas, H., & Mahendra, A. (2017). Machine to machine communication examples and applications. *IoT Worm, 2017*. [online]. Retrieved August 13, 2017, from http://iotworm.com/machine-machine-communication-examples-applications/.
18. Key, C. (2017). 7 Best Developer Tools to Build your Next Internet of Things Application. *Losant*. [Online]. Retrieved July 11, 2017, from https://www.losant.com/blog/7-best-developer-tools-to-build-your-next-internet-of-things-application.
19. Inc, L. (2017). Losant. *Losant*. [Online]. Retrieved July 11, 2017, from https://www.losant.com.
20. IFTTT. Ifttt.com, 2017. [Online]. Retrieved July 11, 2017, from https://ifttt.com.
21. *Wireless: Programmable cellular data, SMS, and voice for devices—Twilio*. Twilio.com. [Online]. Retrieved July 4, 2017, from https://www.twilio.com/wireless.
22. Daniel Minoli, Benedict Occhiogrosso Blockchain mechanisms for IoT security Internet of Things, 2018, 1-2, 1.
23. Rodic-Trmcic, B., Labus, A., Mitrovic, S., Buha, V., & Stanojevic, G. (2017). Internet of things in E-health: An application of wearables in prevention and Well-being. *Emerging trends and applications of the internet of things*, IGI global, pp. 191–197.
24. Rong, K., Hu, G., Lin, Y., Shi, Y., & Liang, G. (2015). Understanding business ecosystem using a 6C framework in internet-of-things-based sectors. *International Journal of Production Economics, 159*, 41–55.
25. Mohamed, N., Al-Jaroodi, J., Lazarova-Molnar, S., Jawhar, I., & Mahmoud, S. (2017). A service-oriented middleware for cloud of things and fog computing supporting smart city applications. In *2017 IEEE smart-world, Ubiquitous Intelligence & Computing, Advanced & Trusted Computed, Scalable Computing & Communications, Cloud & big Data Computing, internet of people and Smart City innovation (smart-world/SCALCOM/UIC/ATC/CBDCom/IOP/SCI)*. IEEE.
26. D. Bruneo, S. Distefano, F. Longo, G. Merlino, A. Puliafito, V. D'Amico, M. Sapienza, G. Torrisi. Stack4Things as a fog computing platform for Smart City applications. In: Computer Communications Workshops (INFOCOM WKSHPS), 2016 IEEE Conference on, IEEE, 2016, pp. 848–853.
27. Ruhaizan Fazrren Ashraff Mohd Nor, Fadhlan H. K. Zaman, Shamry Mubdi. (2017). Smart traffic light for congestion monitoring using Lo RaWAN. *2017 IEEE 8th Control Syst. Grad. Res. Colloquium*, ICSGRC 2017—Proc., no. pp. 132–137.
28. Khatoun, R., & Zeadally, S. (2016). Smart cities: Concepts, architectures, research opportunities. *Communications of the ACM, 59*(8), 46–57.
29. Almeida, R., Oliveira, R., Sousa, D., Luis, M., Senna, C., & Sargento, S. A multi-technology opportunistic platform for environmental data gathering on smart cities. In *Proceedings of the 2017 IEEE Globecom workshops (GC Wkshps)* (pp. 4–8). Singapore: . 2017.
30. Kuzlu, M., Pipattanasomporn, M., & Rahman, S. (2005). Review of communication technologies for smart homes/building applications. *Proceedings of the IEEE Innovative Smart Grid Technologies –Asia (ISGT ASIA)*, IEEE, 2015, pp. 1–6, doi: https://doi.org/10.1109/ISGT-Asia.2015.7437036.
31. Dorri, A., Kanhere, S. S., Jurdak, R., & Gauravaram, P. (2017). Blockchain for IoT security and privacy: The case study of a smart home. In *Proceedings of the IEEE international conference on pervasive computing and communications workshops, PerCom workshops 2017* (pp. 618–623).
32. Retrieved from https://www.apple.com/shop/accessories/all-accessories/homekit
33. Salah, K. (2017). Real time embedded system IPs protection using chaotic maps. In *Ubiquitous computing, electronics and Mobile communication conference (UEMCON), 2017 IEEE 8th annual*. IEEE.
34. Retrieved from https://medium.com/@thewordofsam/how-the-autonomous-car-works-a-technology-overview-5c1ac468606f.
35. Barnaghi et al. (2015). Digital technology adoption in the smart built environment. IET Sector Technical Briefing. The Institution of Engineering and Technology (IET), I. Borthwick (editor).

36. Retrieved from http://industrialiot5g.com/20160714/channels/use-cases/industrial-internet-things-case-study-tag23.
37. Chauduri, S., Rathod, P., & Shaikh, A. (2017). Smart energy meter using Arduino and GSM. *International Conference on Trends in Electronics and Informatics ICEI*.
38. Jayant, P., et al. (2016). Real time energy measurement using smart meter. In *International conference on green engineering and technologies (IC-GET)*.
39. Retrieved from https://www.telit.com/blog/iot-smart-grid-benefits/
40. Chin, J., Callaghan, V., & Ben Allouch, S. (2019). The internet of things: Reflections on the past, present and future from a user centered and smart environments perspective. *Journal of Ambient Intelligence and Smart Environments, 11*(1), 45.
41. Gams, M., Yu-Hua Gu, I., Härmä, A., Muñoz, A., & Tam, V. (2019). Artificial intelligence and ambient intelligence. *Journal of Ambient Intelligence and Smart Environments, 11*(1), 71.
42. Prati, A., Shan, C., & Wang, K. (2019). Sensors, vision and networks: From video surveillance to activity recognition and health monitoring. *Journal of Ambient Intelligence and Smart Environments, 11*(1).
43. Johnston, S. J., Basford, P. J., Bulot, F. M. J., Apetroaie-Cristea, M., Cox, S. J., Loxham, M., & Foster, G. L. (2018). IoT deployment for City scale air quality monitoring with low power wide area networks. In *Proceedigns of the global IoT summit 2018, Bilbao, Spain, 4–7*.
44. Supani, A. (2018). Study on implementation of flood early warning system with internet of things in peri-urban settlement of Palembang for sustainability. *IOP Conf. Series: Earth and Environmental Science, 202*, 1–10.
45. Fiehn, H. B. (2018). Smart agriculture system based on deep learning. *ICSDE'18*, October 18–20, Rabat, Morocco.
46. Mattihalli, C. (2018). Plant leaf diseases detection and auto-medicine. *Internet of Things, 1–2*, 67–73.
47. Jin, D., Hannona, C., Lib, Z., Cortesa, P., Ramarajua, S., Burgessb, P., Buchc, N., & Shahidehpourb, M. (2018). Smart street lighting system: A platform for innovative smart city applications and a new frontier for cyber-security. *The Electricity Journal, 29*, 28–35.
48. Nirosha, K. (2017). Design and implementation of smart movable road divider using IOT. *International Conference on Intelligent Sustainable Systems (ICISS)*.
49. Retrieved from https://chatbotsmagazine.com/chatbots-a-bright-future-in-iot-93fb615b2286.
50. Siddiqui, A. T., & Masud, M. (2017). A system framework for smart class system to boost education and management. *International Journal of Advanced Computer Science and Applications, 7*(10), 102–108. https://doi.org/10.14569/IJACSA.2016.071013.
51. Atabekov, A. (2016). Internet of things-based smart classroom environment: Student research abstract. *Proceedings of the 31st annual ACM symposium on applied computing*, ACM (Apr. 2016), 746–747. doi: https://doi.org/10.1145/2851613.2852011.
52. Chamba-Eras, L., & Aguilar, J. (2017). Augmented reality in a smart classroom—Case study: SaCI. *IEEE Revista Iberoamericana de Tecnologias del Aprendizaje, 12*(4), 165–172. https://doi.org/10.1109/RITA.2017.2776419.
53. Apple. (2015). *Apple invents new apple watch biometric id system using plethysmography*. (cited 25.07.2018). Retrieved from http://www.patentlyapple.com/patently-apple/2015/07/apple-invents-new-apple-watch-biometric-id-system-using-plethysmography.html.
54. Retrieved from http://www.ibm.com/developerworks/library/ba-bluemix-diy-iot-wearable-app/.
55. Tripathy, A. K., Tripathy, P. K., Ray, N. K., & Mohan, S. P. (2018). iTour: The future of smart tourism: An IoT framework for the independent mobility of tourists in smart cities. *IEEE Consumer Electronics Magazine, 7*, 32.
56. Retrieved from https://www.fingent.com/blog/3d-printing-iot-the-future/
57. Retrieved from https://en.wikipedia.org/wiki/3D_scanning
58. Retrieved from https://www.bistasolutions.com/resources/blogs/odoo-iot-erp-disruptive-technology/

IoT Conclusions

Nowadays, IoT is considered one of the cutting-edge technologies in the world. IoT connects everything to the internet. This book discusses all the necessary components and knowledge to start being a vital part of the IoT revolution. IoT is all about intelligence, not just control. Now, IoT is a fast-changing set of technologies and architectures. In this book, we learn how to create smart IoT solutions to help solve problems using Internet of Things (IoT). We presented the most important aspects of IoT, the various applications of IoT, and the enabling technologies for IoT.

This book presents main IoT concepts and abstractions with explaining a case study. Analysis of IoT strength, weakness, opportunities, and threats of IoT is also presented. IoT integrates leading technology such as RFID technology, sensor technology, wireless communication, energy harvesting technology, cloud computing, and advanced internet protocol. IoT is reshaping the world. This book introduces the most important IoT platforms such as AWS, Microsoft Azure, and Watson IoT platform.

In this book, you will understand how IoT is evolving and what the future holds and how it will transform the whole world. Moreover, you will learn the different technologies used in IoT, different IoT hardware platforms, IoT software platforms, IoT cloud platforms, IoT communication and networking platforms.

IoT is now considered a disruptive technology that will shape the future. This book introduces in-depth understanding of IoT from device to cloud and gives valuable insights around different IoT solutions and applications. Moreover, it discusses best practices for accelerating the development of IoT. IoT will enable turning many dreams into reality.

© Springer Nature Switzerland AG 2019
K. S. Mohamed, *The Era of Internet of Things*,
https://doi.org/10.1007/978-3-030-18133-8

Index

© Springer Nature Switzerland AG 2019
K. S. Mohamed, *The Era of Internet of Things*,
https://doi.org/10.1007/978-3-030-18133-8

Printed in the United States
By Bookmasters